James Deane

Ichnographs from the Sandstone of Connecticut River

James Deane

Ichnographs from the Sandstone of Connecticut River

ISBN/EAN: 9783337240950

Printed in Europe, USA, Canada, Australia, Japan

Cover: Foto ©berggeist007 / pixelio.de

More available books at **www.hansebooks.com**

ICHNOGRAPHS

FROM THE

SANDSTONE OF CONNECTICUT RIVER.

BY

JAMES DEANE, M.D.

BOSTON:
LITTLE, BROWN AND COMPANY.
LONDON: S. LOW, SON & CO.
1861.

Entered according to Act of Congress, in the year 1861, by
LITTLE, BROWN AND COMPANY,
In the Clerk's Office of the District Court for the District of Massachusetts.

CAMBRIDGE:
Allen and Farnham, Printers.

INTRODUCTION.

The delineations given in the following pages were made, or directed, by the late Dr. James Deane, of Greenfield, Massachusetts. Living in the immediate vicinity of the quarries whence so many curious traces of animals and of physical phenomena in early geological times have been derived, he had from the first observed them and become deeply interested in them. Scarcely a specimen of any interest was obtained that did not pass under his inspection; and as he beheld, in succession, the gigantic vestige since called *Brontozoum giganteum*, then the smaller ones, gradually descending to those of minute size, the ripple marks, rain drops, trails of insects and worms, presenting themselves, he aspired to divine and determine, if possible, their true nature and relations. To illustrate his own conclusions, and at the same time to afford others enjoying better subsidiary advantages than himself an opportunity to pursue the same line of investigation, and judge of the correctness of his conclusions or substitute their own, he undertook to give delineations, as accurate as they could possibly be made, of the best defined and most characteristic specimens. Some of them were executed on stone, with the most untiring patience, by himself; others were photographed from his selections, and under his direction. He was encouraged in this undertaking by the Smithsonian Institution, which engaged to defray the expense of the drawings and to publish the work in its "Contributions." But in the midst of his enterprise he died, leaving the work, both plates and text, too incomplete to be issued under those auspices. It seemed sad, however, that such exquisite specimens of art, such contributions to science, should be utterly sacrificed. Some of his friends, therefore, felt impelled to collect and arrange, as completely as they were able, what had been done, and, if possible, to obtain from subscribers a sufficient sum to defray the expense of publication at least, and perhaps also to benefit his family. With the coöperation of the Smithsonian Institution, which generously granted the use of the drawings, besides subscribing for a large number of copies, the object has been accomplished, and the work is before us. We believe that these copies, rivalling as they do the actual specimens, will be really useful to those pursuing similar scientific investigations; they will at least furnish a beautiful table-book, to excite an interest in the community in the marvels of nature.

The sketch of his life which follows is an abstract of a Discourse delivered by Dr. Henry I. Bowditch to the Franklin District Medical Society. The compilation of the text, the determination of the species and the additional remarks upon them, and the references to the original specimens, is the work of Thomas T. Bouvé, Esq., who particularly desires to have here expressed his indebtedness to Prof. Hitchcock and to Roswell Field, Esq., for their aid in enabling him to identify many of the stones from which the drawings and photographs were taken, and for other assistance.

AUGUSTUS A. GOULD.

BOSTON, January, 1861.

BIOGRAPHICAL NOTICE.

BY HENRY I. BOWDITCH, M. D.

JAMES DEANE was born February 24, 1801, at Coleraine, Franklin County, Mass. He was the eighth child of Christopher and Prudence Deane. His father was a lineal descendant from James Deane, one of the earliest settlers of Stonington, Conn., whence he removed to Coleraine, soon after his marriage. His home was humble in kind, but placed most magnificently near the summit of one of the highest hills in Franklin County. Monadnock and Wachusett lay immediately within sight, and a few steps from the house enabled the boy to reach a height whence he could, at a glance toward the wide horizon, see all the southern Massachusetts hills. An intense love of nature and beauty seems thus to have been awakened in him from his earliest years.

His father was a hard-working farmer, of a strong mind, and rather puritanic, conservative character. His mother was a woman of sterling piety, good sense, and of a more genial nature. She died when he was about fifteen years of age; and he always cherished her memory with that strength of feeling that every true-hearted son has for a noble mother. Her death made a profound impression on him; and a desire to quit home, and to seek his fortune elsewhere, took possession of him from that hour.

In very early life he attended the district public school during the winter; and subsequently he was allowed to attend, for one term, Deerfield Academy. He was likewise permitted to study Latin for a time, under the direction of Isaac B. Barber, Esq., a lawyer of Coleraine. Later in life he studied French.

If we may believe his own account of himself, he was a clownish youth; but, as one of his schoolmates states, there was a nameless something about him that

BIOGRAPHICAL NOTICE.

caused him to be respected by all his comrades as one superior to themselves, though rather incomprehensible.

When he was about nineteen, convinced that his son would never be content with the life of a farmer, his father consented to let him seek his fortune upon a larger field. Accordingly, with the blessing of his parents, he started for Boston in search of a clerkship, or at least for some position more congenial to his nature. But no path opening to him for employment, he returned, after a few days' absence, to his country home.

At the age of twenty-one he took a final leave of his home, without a penny in his pocket, but with a brave, manly, honest heart beating warmly and hopefully in his bosom. He went to Greenfield, and offered himself as clerk to Elijah Alvord, Esq., then Clerk of the Court and Register of Probate. Fortunate beyond expression was the poor youth in meeting this excellent man. Mr. Alvord seems early to have appreciated the many high qualities possessed by his young assistant. He seems, moreover, to have felt more than most persons do, the responsibility that was imposed upon himself to aid such a man in his career. Mr. Deane was received into the family, and there resided for four years. These were some of the happiest in his life. Mr. and Mrs. Alvord treated him like a son. His heart and intellect expanded under the warm influences of a kindly sympathy, and a sense of gratification in the performance of more pleasant and more profitable duties. The same unobtrusive deportment, and an entire faithfulness in the performance of every duty, with a rich vein of genial humor underlying all his actions, marked his career. The emolument was small; but with it he was enabled not only to aid his parents, but to pay for the education of a younger sister.

During the latter part of his engagement with Mr. Alvord, that gentleman permitted him, while still a clerk in the office, to become a pupil of Dr. Brigham, at that time an eminent practitioner in Greenfield, and to spend a part of each day in study. In 1829–1830 he attended his first course of Lectures in New York, given by the well known and able Professors Delafield, Stevens, Smith, Beck, and their associates.

He received his degree of Doctor of Medicine in March, 1831, and soon afterwards commenced the practice of his profession in Greenfield. He had no introduction save his own character and mind, and those who knew him are well aware that no man was ever less a trumpeter of his own fame. Many physicians entered the town at this, and at subsequent periods, but almost all retired before his death, long before which time his own success had become complete. I think it may be said, not invidiously, that he took the first rank as a surgeon in that vicinity. For many years he experienced the bitter influences of poverty. But though straitened in means, he would never consent to become a mere routinist in the profession. He knew that there was progress, and he determined not to fall behind the foremost. Accordingly, in 1849, he quitted practice and spent several weeks in study at New York with immense advantage.

In 1837, six years after beginning practice, he sent a communication to the Boston Medical and Surgical Journal. From that time until January, 1855, he was a frequent contributor to the pages of that journal. These papers are all written in a curt, pithy style, exactly to the point, with not a word too many or too few. He evidently never writes for mere effect, but simply to tell, as clearly and as concisely as possible, whatever he meets with in his daily practice that he thinks will be of real value to his profession. The papers are mostly on the surgical cases, but he likewise records some very interesting cases in medicine proper and pathology. We can trace the gradual rising of his reputation by the gradually increasing severity of the accidents mentioned. These records, by their gentle allusions and occasional bursts of real feeling, demonstrate that he never operated without having a feminine tenderness for the suffering of his patients.

In February, 1854, he delivered an address before the Franklin District Medical Society, admirable in its philosophy, and wise in the accumulated experience of seventeen years of active practice. The subject of this address was "Fractures of the Thigh."

In May, 1855, he prepared, at the request of a committee of the Massachusetts Medical Society, a communication on the "Hygienic Condition of the Survivors of Ovariotomy."

In 1854 he was chosen Vice-President of the Massachusetts Medical Society, which office he held for the usual period of two years.

We have alluded to Dr. Deane's delight in nature. Even while a mere child he made his juvenile scientific investigations upon the growth of trees, and spent his time much more eagerly in the woods and along the trout brooks than in the milking of cows.

In the early part of the year 1835, with many of his fellow-citizens, he noticed some curious impressions upon slabs of stratified sandstone that had been brought from Turner's Falls, near Greenfield, to be used for sidewalks in the village. One of his wisest of his fellow townsmen remarked, "We all saw them, and, mentally at least, exclaimed, 'they are bird-tracks,' and then went away and thought no more about them." Dr. Deane alone recognized in their mute teachings sublime indications of an Almighty hand. He seems from that moment to have entered upon the investigation of the whole subject with that earnest but quiet and never yielding enthusiasm, that had always been his peculiar characteristic. He sought knowledge everywhere upon the subject. He visited the spot whence the slabs had been procured. He wrote to the men most noted for their learning in geology. At first, the whole scientific world was against his belief that they were the tracks of birds. Gradually the naturalists of this country, and many in Europe, yielded to the accumulated evidence, and more especially to the facts as subsequently given to the public by Prof. Hitchcock in a scientific form; and the tracks became generally regarded as those of birds. There is reason to believe, however, that before his death, Dr. Deane materially modified his opinion on this point. In proof of it we could refer to conversations held with Mr. Roswell Field and others, a few months before his decease, and to letters to Drs. Gould and Wyman, which will be quoted in the text. It is not, however, upon the precise point as to whether these tracks were those of birds or otherwise, that we dwell. The object we have in view is to show that from the first moment that the tracks came under his observation they were ever the source of deep interest and severe study for him. Particularly was this the case during the last few years and months of his life, when every moment that could be spared from great professional labor was directed to this his darling study. In proof of this, let the following facts and statements be noted.

In 1844 he sent a paper to the American Journal of Science, which was published, with drawings; and he remarks that it is "to accumulate facts that bear upon this interesting subject" that he presents the communication.

In 1845 (vol. 49, p. 79) he describes other species of tracks, probably those of a batrachian reptile. He closes with these words: "An indescribable interest is imparted by opening the long-sealed volume that contains the records of these extinct animals. The slabs were uncovered and raised under my supervision, and page after page, with their living inscriptions, revealed living truths. There were the characters, fresh as upon the morning when they were impressed; reminding the spectator of the brevity of human antiquity, and of the frail tenure of human works. On that morning, how long ago no one can tell or will ever know, gentle showers watered the earth, an ocean was unruffled, and upon its borders primæval beings enjoyed their existence, and inscribed their eventful history." In a more elaborate paper (vol. 48, p. 158) he describes new discoveries, a stupendous impression of a foot half a yard long, and capable of holding half a gallon of water! He adds: "What was the real magnitude of this powerful bird? He maintained his supremacy throughout the entire period of the new red sandstone deposition, while other varieties, though gigantic and powerful, became extinct. He was endowed with a physical frame fitted to endure the turbulence of the era in which he reigned supreme monarch of his race, and was finally exterminated only by the all-pervading catastrophes that swept from the earth other vast creatures which were his contemporaries, but not his conquerors."

In 1847 he describes the track of a quadruped, being the fourth that had been discovered. He infers that these early inhabitants of this planet frequented regions which were periodically, or at least occasionally, submerged. He discusses the causes, and thinks these inundations arose from sudden floods.

Finally, in 1848 (January), he gives a very brief account of another quadruped track.

Meanwhile, he had sent some specimens and a letter, dated Sept. 20, 1842, to Dr. Mantell in England. This gentleman laid the communication and the tracks before the Geological Society of London. The previous scepticism of Prof. Owen and of other eminent geologists and palæontologists was thoroughly overcome by this communication, conjoined, as it was, with the discovery of the Dinornis of New Zealand. Dr. Mantell remarks, "Your brief and lucid description," with the specimens, has placed this subject before the geologists of England in a "most clear and satisfactory light;" and "the thanks of the Society were warmly and unanimously expressed for so valuable a communication."

In 1849 a still more elaborate memoir, with many plates, was sent to the American Academy of Arts and Sciences, and was published in the Memoirs of the Academy, Vol. IV. New Series.

Two similar papers were published, in 1850 and 1856, by the Academy of Natural Sciences at Philadelphia; and in one of these he first describes the minute tracks ascribed to insects (Vol. II. 7, and Vol. III. 173).

It will not be here attempted to give an analysis of these papers. Suffice it to say, that in them he alludes not only to bird-tracks, but to impressions made by many different kinds of the lower animals, even to those of insects and crustaceans.

During all the time he was preparing these papers he was constantly making drawings of new specimens, hoping that at some future day the complete whole might be published as the crowning labor of his life. Nothing could be more touching than the quiet but determined manner with which he went on daily accumulating his facts. Utterly unable to see how, with his small means, the work could ever see the light, he still struggled on in a sublime faith. The amount of labor he performed seems quite marvellous when we remember that he was constantly engaged in an extensive practice, which spared to him no certain hours of study. Called hither and thither over an extent of twenty or thirty miles radius, surrounded by quackery that gnawed at and traduced him; conscious of his powers, yet morbidly sensitive to the idea that he was not duly appreciated by some whom he respected, it was, doubtless, with a sense of divine consolation that he turned to these relics of a past era, and with a generous ambition labored to present them to his fellows. As LaGrange of old sought "in his peaceful mathematics" a relief from the world, so our friend found, in this beautiful study, a never-failing resource from the corroding cares of earth.

In 1836, Dr. Deane married Miss Mary Clapp Russell, of Greenfield. He was eminently a domestic man, a most tender husband and loving parent. His townsmen will long remember his upright, manly intercourse with them as neighbor, friend, and physician. Though a man of few words, he was eminently genial and social. He possessed to a remarkable degree a love of fun and a power of mimicry, ordinarily masked, however, by the graver tones of his character. In his political views and actions he was clear and reliable. Without offensively thrusting his opinions upon any one, he was no coward in the utterance of any sentiment he thought right. Throughout adult life he was a consistent and fearless defender of the rights of man. His taste for the sister arts of music and painting was admirable and wholly natural, for he had no instructors. From boyhood, he used his pencil and his pen in free but extraordinarily accurate sketches. We hazard nothing in saying, that, had he chosen either music or drawing as a profession, he must necessarily have taken a first rank among the professors of those twin handmaids of Beauty. He has left some few rhythmical attempts, *torsos*, so to speak, in the divine art of Poesy.

He received the honorary degree of Master of Arts from Amherst College in 1838, and was a Corresponding Member of Natural History Societies of Montreal and Boston.

In his professional relations he was exemplary in all points of etiquette, but it is to be feared, that, holding as he did the profession of medicine in the highest esteem, he was morbidly sensitive to the support of quackery evinced by some in whom he thought to see a wiser course of conduct. He forgot that while credulity and hope remain in the human heart on one side, and craftiness or folly exists on the other, quackery will always flourish. In his religious views he was simple and true, but his precise opinions in regard to specific doctrines were not formally avowed. This much may be asserted of him: he believed that no *profession* compared with a *life of goodness*. His reverence and love of God were unbounded. He had always lived among the hills, surrounded by the beauty of God's works, and his soul bowed before him as the beneficent Creator and loving Father.

In reviewing the prominent facts of Dr. Deane's life, we find him, as a child of poor but religious parents, drinking in, with every sense, all the beauty and grandeur of nature so lavishly spread before him. Early in life, and perhaps unconsciously, he makes his protest against mere physical labor, and devotes himself to the cultivation of his intellect and his tastes. He grows slowly, without show or special elegance, but with a solid firmness. Turning readily to that noble profession which is so capable of enriching the mind and expanding the heart of its votary, he reaches the highest point of reputation with his peers, his professional associates. During these professional engagements in the daily walks of life, he still keeps his eyes open to the revelations of nature. Horticulture is his pleasant recreation, and a drive among the hills gives him infinite delight. Suddenly he awakes to a new existence in the recognition of the sublime significance wrapped up in a simple footstone near his dwelling. His highest, his religious nature, is ever afterwards constantly appealed to, while bringing to light specimen after specimen of these relics of a past age. Each part of his life seems fitted "by natural piety" to all the others. With the exception derived from the inherent imperfections of our nature, we see a beautiful and symmetrical whole, a well-proportioned, stalwart intellect, guided by an honest, earnest, religious soul.

Dr. Deane was of a tall and commanding form, half a head taller than most men, and of a well-knit, compact frame. His very walk conveyed an idea of strength. Having enjoyed more health than usually falls to the lot of man, he for almost the first time in his life was seriously ill in 1853. Though he recovered completely, this illness seems to have taught him the frail tenure of his life. During the spring of 1858 he began to suffer from headaches, and he was less well generally. Finally, about three weeks before his death, he was struck down by a typhoidal series of symptoms, which gradually augmented until he quietly sunk into his last sleep, June 8, 1858, amid the universal sorrows of the community in which he lived.

LIST OF PUBLISHED PAPERS.

In the Boston Medical and Surgical Journal.

1. Congenital Fissures of the Palate. Vol. 16, page 333, June 28, 1837.
2. Extraordinary case of Spasms of the Voluntary Muscles. Vol. 28, 336. 1843.
3. Polypus in Utero, of unusual size. Vol. 30, p. 449. July, 1844.
4. Case of Carditis; very curious; allusions, likewise, to the treatment pursued in a case of Obstinate Constipation. Vol. 32, p. 158. 1845.
5. Iodine Injections in the Treatment of Hydrocele, etc. Vol. 33, p. 18. 1845.
6. Inhalation of Ether; cases. Vol. 37, p. 18. December, 1847.
7. Fibrous Tumors of the Uterus; Gastrotomy; very interesting, as he was obliged to close the wound without removal of tumor; recovery of patient. Vol. 32, p. 221. October, 1848.
8. Abscess of the Tibia; trephining of the bone. Vol. 43, p. 131. September, 1856.
9. Ovariotomy; cases; practical suggestions. Vol. 44, p. 474. 1851.
10. Lithotripsy in a Female. Vol. 46, p. 20. 1852.
11. Tumors of Various Kinds; diagnosis, etc. Vol. 46, p. 400. 1852.
12. Chronic Laryngitis; case. Vol. 46, p. 503. 1853.
13. Popliteal Aneurism; operations. Vol. 48, p. 141. June, 1853.
14. Union of Divided Fingers; cases. Vol. 48, p. 329. May, 1853.
15. Immense Tumor of the Parotid; operation; radical cure. Vol. 48, p. 486. January, 1854.
16. Fractures of the Femur and their Treatment; an address delivered before the Franklin District Medical Society, at their meeting in February, 1854. Vol. 50, p. 57.
17. Polypus of the Uterus. Vol. 51, p. 141. September, 1854.
18. Ligature of the Femoral Artery; important case. Vol. 53, p. 1. August, 1855.
19. Case of Osteo-Aneurism; very interesting. Vol. 53, p. 444. December, 1855.

LIST OF PUBLISHED PAPERS.

In the Proceedings of the Massachusetts Medical Society.

1. On the Hygienic Condition of the Survivors of Ovariotomy. It was likewise republished by John Wilson & Co., Boston.

In Silliman's Journal of Science.

1. Letter to Dr. Mantell, etc. Vol. 45, p. 178. October, 1843.
2. On Fossil Footmarks of Turner's Falls; plate. Vol. 46, p. 73. April, 1844.
3. On the Discovery of the Fossil Footmarks. Vol. 47, p. 292. October, 1844.
4. Fossil Prints in the New Red Sandstone of the Connecticut Valley; Batrachian-Reptile or Marsupial-Mammalian tracks; rain drops, with plate. Vol. 48, p. 158. 1845.
5. Extract from a Letter to Prof. Silliman. Vol. 49, p. 213. 1845.
6. Notices of New Fossil Footprints. Vol. 3, New Series, p. 75. January, 1847.
7. Footprints of a New Species of Quadruped. Vol. 5, p. 40. 1848.

In the Memoirs of the American Academy of Arts and Sciences.

1. Illustrations of Fossil Footprints of the Valley of the Connecticut; with nine plates. Vol. 4, New Series, p. 209. 1849.

In the Journal of the Academy of Natural Sciences, Philadelphia.

1. Fossil Footprints of Connecticut River. Vol. 2, Second Series, p. 71. 1850-1854.
2. On Sandstone Fossils of Connecticut River. Vol. 3, Part 3, p. 173.

A

MEMOIR

UPON

THE FOSSIL FOOTPRINTS AND OTHER IMPRESSIONS

OF THE

CONNECTICUT RIVER SANDSTONE.

BY

JAMES DEANE, M. D.

The compiler of the incomplete papers left by Dr. Deane has not deemed it desirable that he himself should undertake to describe, scientifically, such impressions as Dr. Deane left no particular description of, especially as the recently published work of Dr. Hitchcock, upon the Ichnology of Massachusetts, embraces an account of perhaps all the species. He has therefore limited his own attempts in the notices of the Plates, to obtaining references to the stones from which they were taken, and the cabinets in which these stones are now to be found; to a determination of the species, and the works in which they are described; and more especially, to their identification with species given in the Ichnology of Massachusetts; adding such other matter of general character as he judged might be of service.

The colored lithographs are in the same style as those given by Isaac Lea, Esq., of Philadelphia, accompanying his Memoir on the Sauropus primævus, and which were much admired. They represent well the prevailing tint of the sandstone. The drawings which were made by Dr. Deane on stone, it may be confidently stated, can never be surpassed. Their accuracy makes the possession of the Plates almost as valuable, even for scientific descriptions, as the original specimens.

The nine Plates, 16, 17, 31, 40, 41, and 43 to 46 inclusive, which it was intended to give in photo-lithographs, are direct photographs from the original stones, and are exquisite specimens of art.

T. T. BOUVÉ.

MEMOIR.

The object of the following memoir is to illustrate the sandstone footprints of Connecticut River. The broad range of animal organization suggested by these expressive vestiges, and the remote geological antiquity of the formation in which they occur, are considerations that invest them with an extraordinary degree of scientific interest. Prior to their discovery the sandstone rock was considered to be quite barren of the indications of animal life; but, in place of its former insignificance, it is now known to be exceedingly rich in the evidences of organic existence.

Literary History of the Footprints. — The earliest written descriptions of the footprints were communicated by the author to several gentlemen of scientific eminence, in the year 1835. In these original accounts the impressions were described as belonging to birds, and the probable manner of their occurrence explained. The first published account appeared in the Journal of Arts and Sciences for 1836, from the pen of Dr. Hitchcock. Other occasional papers by this gentleman were subsequently communicated through the same channel. A more complete treatise was published by Dr. Hitchcock, in his Final Report to the Legislature of Massachusetts, in 1842, and another in the third volume of the Transactions of the American Academy of Arts and Sciences. Several papers were also presented by him to the American Association for the Advancement of Science at the annual sessions of that body.

Single papers, descriptive of sandstone footprints, have been communicated to the public by several gentlemen. By William C. Redfield, Esq. (Am. Jour. of Sci. vol. 34, p. 134); by Dr. John C. Warren (pamphlet); by Dr. Alfred T. King (Proc. Acad. of Nat. Sci.); by Isaac Lea, Esq., several illustrated papers upon footprints and bones from the New Red Sandstone of Pennsylvania (Proc. Am. Phil. Soc. 1849); and by Charles Lyell, Esq., upon Footprints of Birds and Impressions of Rain-drops in the Valley of the Connecticut River, communicated to the London Geological Society.

The descriptive papers hitherto published by the author have been contributed to various numbers of Silliman's Journal; to the Journals of the Society of Natural History, Boston; the Academy of Natural Sciences, Philadelphia; and to the Academy of Arts and Sciences, Boston.

This list, so far as I am able to ascertain, includes the principal contributors to the literature of the sandstone footprints of this country. It is also proper to allude to the successful labors of those gentlemen who have quietly pursued their investigations in the field without recording the results of their observations. Among the most conspicuous was the late Mr. Marsh of Greenfield, who accumulated a series of footprints of great value to science. Roswell Field, Esq., a gentleman of acute powers of observation, succeeded Mr. Marsh as an explorer; and, possessing an intimate knowledge of the subject, began at once to make discoveries of significant importance. His estate, being at Turner's Falls, embraces the richest localities yet discovered, and his success has been very remarkable. In the preparation of this paper the author will have frequent occasions to acknowledge his obligations to him for original materials and observations.

Analogies of the Footprints. — It is proper, for a clear comprehension of this topic, to premise, that the footprints will be assumed to indicate animals whose representatives are included in the existing classes of Birds, Reptiles, and Insects; and there are also presumptive proofs that the Mammalian class also coëxisted. The birds may, with probability, be subdivided into two principal orders, — the waders proper, and the brevipennate or struthious tribes. They were usually tridactylous, but a portion of them possessed a fourth or rudimentary toe, that performed no part in the office of walking. Without exception there are three toes pointing forward, and there are never more than four toes. Each of these divisions of the ornithoid footprints included a great variety and combination of forms, some of which continued to be produced throughout the entire deposition of the sandstone rock, while others successively disappeared and were replaced by new and distinct creations.

The reptilian footprints, although numerous, bear but a small proportion to those of the birds, in number or variety. A portion of this class of footprints appears to indicate animals intimately related to existing saurian reptiles; but another portion of them, comprising several distinct varieties, cannot be compared with any known type. They are both quadrupedal and bipedal, and, considered as a class, possess extraordinary interest.

The structural organizations of the extinct animals can only be inferred by applying the laws of comparison to the impress of their feet; for the configuration of the footprint presents the only practical basis for comparison, and it fortunately happens that impressions occur so accurately defined as to supply, in an eminent degree, an equivalent for the missing bones of the animal to whom they are due. It would seem that the exact impress of the foot offers a basis of analogy little inferior in value to the foot itself. We instinctively judge of the physical organizations of familiar animals by their footprints; and the difficulty in comprehending the organisms of the extinct animals lies, in a great degree, in an insufficient acquaintance with the footprints of their living successors. If living animals be found whose footprints conform in every essential particular to the fossil impression, it must in reason be conceded, that the organization and habits of the extinct and living types are also conformable.

The comparison of the supposed ornithoid vestiges with the footprints of living birds is unequivocal; a remarkable agreement exists between them; but in the reptilian impressions, that do not usually exhibit the phalangeal divisions of the toes, the grounds of analogy are not so clear. There are, moreover, examples of this class of footprints, of a character so anomalous, as, in the present state of science, to defy comparison. Certain forms of the footprints (Pl. 37) indicate bipedal monsters, that, in respect to magnitude and intricate mechanism of the feet, are without analogy in existing nature. The original from which this Plate is reduced, is seventeen inches in length and eleven in breadth, and its impress is without blemish. Other bipedal reptiles existed, that were distinguished for their diminutive proportions, whose footprints are represented in Plates 38 and 39. A formidable obstacle to comparison exists in the fact, that a certain portion of the extinct animals appears to have been invested with diversified powers of locomotion. It is presumed they could walk as quadrupeds (Pls. 33 and 34) or jump like the kangaroo (Pls. 31 and 32); and it may hereafter be inferred that they could also walk upon their posterior feet as bipeds. If this inference be sustained, we may look in vain for living representatives whose powers of locomotion present such extraordinary combinations.

The ornithic footprints are, as has been remarked, essentially tridactylous; a small proportion of the waders only, exhibit the fourth toe, which is fixed upon the metatarsal bone at a point above the plane of the anterior feet. It is directed backward and inward as in the living tribes, and in some instances the extremity only of its nail touches the earth (Pl. 5). Its impress is rarely seen except in those examples of footprints wherein the bird sank deeply in the soft materials of the stratum, bringing it down to the level of the surface.

An identical system of articulation of the toes is common to all the ornithoid footprints; and this system is also identical with the corresponding organs of living birds. The number and arrangement of the digital phalanges are uniformly the same in each type of birds, and there is a like agreement in the tarsal and ungual systems; and this uniform correspondence constitutes the basis upon which the affinities and relations of the extinct animals can be comprehended. In both extinct and living races the inner toe is invariably shortest, the middle longest, while the outer holds an intermediate grade. There is no exception to this rule among the fossil footprints. In existing tridactylous birds there are three phalanges of bones for the inner toe, four for the middle, and five for the outer; but as the terminal phalanx belongs exclusively to the nail, there are in the footprint two concave depressions for the inner toe, three for the middle, and four for the outer, that correspond to the tuberous expansions of the respective articulations (Pl. 1, figs. 2 and 3, and Pl. 2, figs. 1, 2, and 3, impressions of living birds, and Pl. 11 and others, fossil impressions). There are, in addition to the phalangeal and ungual markings in well-defined examples of footprints, a group of three concavities, situated posteriorly to the toes (Pl. 4, fig. 1), that are impressed by the tubercles of the metatarsal bones that support the respective toes. There is usually considerable variation in these features of the footprint; they are wanting altogether in the digitigrade examples, and in Plate 7 they are blended so as to destroy their outlines. The central tubercle, or that corresponding to the middle toe, is also modified by compression of the first phalanges of the respective toes, that embrace it (Pl. 10), and change its form. It is also unusual for the tubercle supporting the inner toe to be impressed at all (Pl. 4, fig. 4), and in some forms of the struthoid examples the impress of the tubercle supporting the outer toe is very massive and preponderating (Pls. 9, 11, and 13). The nails terminating the toes are broad at their contact with the last phalanx, and are stout and blunt; and, in some very rare instances, the dermoid papillous elevations are distinctly impressed (Pl. 16).

In these typical features of the ornithic footprints, it will be seen that there is an exact agreement with the footprints of living tridactylous birds, that apparently establishes the cognate relations of the extinct and living races. The same alternation and consecutive series of steps distinguish the movements of each type of birds; and, in all essential particulars, a mutual agreement exists, that appears to connect them by indubitable analogies. Until a recent period no fact has ever appeared to disturb the ornithic theory of a portion of the footprints; but late discoveries disclose the singular fact, that other contemporaneous animals that were quadrupedal possessed a similar organization of the posterior foot (Pls. 33 and 34). But, from the accurate comparison of the ornithoid impressions with those of existing birds, it is proper to consider them to be analogous, although future discoveries may prove the grounds of this belief to be fallacious.

The reptilian footprints, that are supposed to indicate animals of saurian type, are distinguished for inferior magnitude and for the disproportion of the anterior and posterior feet, the latter being several times greater than the former (Pls. 23 to 26 incl.). This characteristic feature is indeed common to all varieties of quadrupedal footprints. As a general rule each foot consists of four toes; but the anterior foot sometimes has five (Pl. 25, fig. 1). The toes are comparatively stout, and the feet massive. In some examples the toes are in contact (Pl. 23), but they usually radiate from a common centre (Pl. 24). In other examples, those of the posterior feet are parallel, while those of the anterior feet radiate (Pl. 23). With one exception (Pl. 24, fig. 2), each foot points obliquely outward; and the divergence of the anterior foot is sometimes so great as to point directly outward (Pl. 23). As an *invariable* rule the ornithic footprint inclines inward, and the inclination is in direct proportion to the length of stride. When the step is long, the convergence of the foot is slight; but when it is short, the foot points prominently inward. The central toe, in ornithoid footprints, points with precision to the succeeding footprint. The convergence or divergence of the footprints, then, indicates the class to which the animals making them belonged. It rarely happens that the phalangeal divisions of the toes, in the saurian types, can be determined, although it is sometimes the case; yet the symmetrical proportions of their feet are usually well preserved. In walking, the posterior foot falls upon the ground near the place of the anterior foot, usually a little distance behind (Pl. 24, fig. 2), within and behind (Pl. 24, fig. 1), before and without

(Pl. 25, fig. 3), without (Pl. 26, fig. 1), and sometimes the impress of the anterior foot is partially or altogether obliterated by that of the posterior foot, and both pairs of the feet fall in regular alternation.

The type of animals indicated by these diminutive footprints may be referred with probability to the saurian order of existing reptiles, for there is an intimate analogy in the structural configuration of their feet. Plate 21 represents the footprints of a living alligator, and by comparing them with those of Plate 24, fig. 1, the points of resemblance will be seen to be very intimate. The posterior feet in each Plate have the same number, form, and radiation of the toes, and the three inner ones in both are pointed with claws; and there are also like impressions of the metatarsi. The form and radiation of the anterior feet are also conformable; and upon these analogies a portion of these quadrupedal impressions will be assumed to indicate animals who were the archetypes of certain existing saurian reptiles.

The impressions upon Plate 29, which are very minute, indicate a different tribe of quadrupedal animals. Their feet were widely separated, and the anterior foot is planted considerably in advance of the posterior.

The origin of the remaining quadrupedal footprints cannot be clearly inferred upon any known existing analogies; and the interest created by an investigation of them is in proportion to their obscurity and anomalous combinations. The footprints upon Plates 31 to 34 inclusive, indicate animals whose powers of locomotion were diversified, and who approached more nearly to the mammalian type than any other form of the sandstone footprints. It will be seen in the analysis of these remarkable impressions, that the feet partake of the characteristics of both birds and reptiles; that is, the posterior foot conforms strictly to the ornithic impressions, and the anterior to the reptilian. In Plates 33 and 34 the larger feet are, in respect to the phalangeal arrangement, the heel and nails, conformable to the ornithic impressions; while the lesser, or anterior footprints, are strictly upon the reptilian type. In this style of quadrupedal footprints, the same alternation of the feet occurs as in the foregoing saurian examples: the anterior and posterior feet fall upon the surface nearly together; and, if the posterior footprint be considered independently of its companion, it will without hesitation be pronounced to be ornithic in its character and indications.

In Plates 31 and 32 the same type of feet prevails, but their positions and relations are materially different, as if the creature had assumed the jumping movement for the walking. The smaller, or anterior footprints, are arranged in pairs (Pl. 32), at the right of the Plate, and are so perfectly defined as to exhibit the phalangeal divisions of the toe. The posterior or ornithic feet will readily be recognized; but in addition to those of the preceding examples (Pls. 33 and 34), the tarsus is prolonged backward, so as to present its entire and unblemished impress. The entire group of impressions was produced by the animal when in a sitting posture; but, excepting distinctions depending upon a difference of species, they appear to be essentially identical with those of Plates 33 and 34, one set being impressed in the act of walking and the other in leaping. The singular resemblance of the posterior feet in these remarkable forms of the quadrupedal footprints to those of the ornithic varieties, is a circumstance of significant meaning. Associated as they are with footprints of unquestionable reptilian type, the first fact appears that throws a doubt upon the subject of the *ornithic* origin of the footprints. If it shall be proved by future discoveries that the animals making these complicated impressions possessed the additional power of walking upon their posterior feet alone, the ornithic theory of the footprints would be settled in a summary manner, impregnable as it now seems to be.

Another modification of the ornithic-reptilian footprints appears upon Plate 35. It might appear, by the wide separation of the right and left sets of feet, and from the dragging of the toes in all the feet from step to step, that the animal was created upon some testudinal type, but the character of the posterior foot forbids this inference; and, moreover, there is no evidence that tortoises existed during the deposition of the sandstone formation, although a different opinion has been formerly held.

The footprints upon Plates 36, 37, 38, and 39 are those of bipedal reptiles, and their origin is involved in profound obscurity; those upon Plate 36 are remarkable for the form of the feet, and for being associated with the impression of a dragging tail; those upon Plates 38 and 39 are distinguished for their diminutive size, while that upon Plate 37 is distinguished for its enormous magnitude. It is the largest, the most massive and solid, footprint ever discovered, the original being seventeen inches in length and eleven in breadth. These stupendous but graphic impressions cannot be contemplated without astonishment; but it is vain to assign the creature who made them a place in the scale of animal organization.

It will now be understood, that the footprints indicate by their analogies animals whose prototypes are to be found in the existing classes of Birds and Reptiles. Impressions of Fishes also occur abundantly at Turner's Falls, in beds alternating with the footprints. While, therefore, the shores of the ancient sandstone basin were thronged with air-breathing, warm-blooded animals, its waters teemed with fishes and amphibious creatures.

Next, and lastly, in this interesting drama of life we have the class Insecta; and of all the fossils in the sandstone strata there are none that excite admiration and wonder so much as these delicate vestiges of insects. Their impress is unblemished. The gigantic footprints confound the imagination by their immensity; but these inimitable inscriptions interest us by their surprising delicacy and beauty (Plates 40, 41, and 42).

Associated with these attenuated footprints are the trailing impressions of larvæ, which are abundant in most localities. There are also other markings upon the sandstone strata resulting from organization, the meaning of which is incomprehensible. The fossils delineated upon Plates 43 and 44 are of this description. It will require long years of patient investigation to solve the mysteries that are written in the sandstone language. Even the falling rain has registered its duration and intensity upon the impressible strata.

It is unfortunate, for a satisfactory solution of the obscure origin of the footprints, that the osseous systems of the animals making them cannot be recognized. Bones have been occasionally found; but in a condition so imperfect as to exclude legitimate conclusions. They are, however, uniformly hollow, and are filled with the same material as the rock in which they occur. That the skeleton should be associated with footprints, it is indispensable that the animals should perish upon the marginal grounds of the sandstone basin, which were periodically submerged, and that their bodies should be immediately buried by an overspread of the materials of the succeeding stratum. But the cylindrical, permeable bones of birds would cause the carcass to be lifted and floated away by the retreating waters, and hence it would be deposited in places remote from the stratified division of the rock. The discovery of the skeleton, therefore, cannot be confidently expected in the fossil-bearing strata.

Footprints in situ.—When the footprints are contemplated in their original situation, their importance, as indicative of events that transpired in a remote geological era, is strikingly manifest. The sandstone strata have been elevated since their deposition, and their original level positions transformed into various angles of elevation. The fossil-bearing strata present smooth, lustrous surfaces,— a condition that results from the precipitation upon the coarser materials of the strata of such fine argillaceous particles as were held in suspension by the agitated waters; and upon this polished film, when exposed by the retreating flood, the animals impressed their footsteps as upon wax. The plastic and retentive properties of this sedimentary deposit were very remarkable, the minutest details of organization being engraved upon it sharp as the impress of a die.

When several contiguous strata are impressed with footprints, each succeeding stratum bears upon its inferior surface an exact copy of the stratum beneath; and hence it happens that certain strata bear intagliated impressions upon their superior, and relievo casts upon their inferior, surfaces. When a new layer overspreads a preceding deposit, an interval of quiet and sunshine succeeded; by parting with its water its surface became consolidated, and in this condition the animals imparted to it the impress of their feet, which has been preserved unchanged by the succeeding overflow of plastic mud, and thus the process continued indefinitely, each stratum not only taking the form of its predecessor, but preserving its records from destruction.

The original deposition of the sandstone beds covered an area of considerable length, but of limited breadth. Its present northern terminus is at a cataract in the Connecticut River known as Miller's Falls, near the northern boundary of Massachusetts, and, stretching to the south, it intersects this State and also that of Connecticut. The Connecticut River proceeds, in its southern course, directly over the sandstone beds, as far as Middletown, in the latter State, and is thence diverted to the east, over beds of metamorphic rocks, the sandstone continuing in its straight course, and terminating at the Long Island Sound, at New Haven. Footprints usually occur wherever the stratified division of the rock is extensively quarried, as at Middletown and Weathersfield in Connecticut; and at Chicopee, South Hadley Falls, and Turner's Falls, and at other localities in Massachusetts. The localities are numerous, and are distributed over an extent of more than one hundred miles. The facilities for investigation are in proportion to the elevation of the original beds; but footprints do not occur in all situations in the same degree of perfection. At the cataracts of Turner's Falls they are not only most abundant, but of extraordinary excellence. Their superiority perfectly adapts them to the purposes of illustration, and the originals of the Plates have been selected from this prolific field. The region of Turner's Falls, constituting the northern limit of the sandstone beds, appears to have been a common rendezvous of these ancient animals; while the retentive properties of the strata were singularly favorable to the preservation of their footprints. So numerous were the creatures that resorted to this attractive region, that the strata are often literally covered with their footsteps. When, however, they are thus associated in numbers, it is manifest that they were not all simultaneously impressed. Some of them are deep and imperfect; others are superficial; and when the materials of the strata became, by exposure, too dense to receive articular forms, all that resulted from the transits of the birds is the indentation of their blunt claws.

The stratified beds of sandstone were evidently deposited upon the rim or marginal limits of the ancient basin or receptacle of water, the materials being transported to the place of deposition by the agency of powerful floods or streams. The strata were of course deposited in a slightly inclined position, and by perpetual succession acquired a depth of many thousand feet, and throughout, footprints occur at irregular intervals, even at the bottom of the sands. At the commencement of the sandstone deposition the animals whose remarkable history it commemorates were already in existence, and they flourished, without intermission, to its completion.

It is usual to find several contiguous fossil-bearing strata that are separated by wide intervals of a coarser description of material and stratification, that contain no evidences of organic life. The proportion of the latter to the former is very insignificant; and hence the detection of the footprints is uncertain, and depends as much upon chance as upon exploring discernment. Their search, however, is much facilitated by negative considerations: it is not difficult to determine the lithological conditions under which they *do not* occur. When the rock is coarse or granular or conglomerated, and is destitute of that delicate glazed deposit that results from earthy precipitation, it is in vain to search for footprints; but when the stratification approaches the condition of shales, and their surfaces are smooth and bright, it may be inferred that footprints exist. The fossil strata are comparatively thin, being at most a few inches only in thickness.

The elevation of the sandstone beds has been accomplished by an irruption of basaltic rock in a melted condition; and in many places at the outbreak of the igneous rock, the sedimentary beds are thrown up into hills of considerable elevation. In some instances the elevation has been effected without material change of the natural inclination of the strata. The irruption of the basaltic rock is coextensive with the sedimentary formation, commencing with it at its northern terminus, and accompanying it throughout in low, bleak ridges, throwing up, in eastern or upper aspect, the fossil-bearing strata.

Causes that have prevented the perfect Preservation of the Footprints. — Several circumstances have interfered to prevent the preservation of the footprints, the most prominent being that which depends upon the variable density of the stratum at the time of the animal's passage over it. If the foot was planted upon the stratum while yet too soft to retain its form, the impress was of necessity obliterated, or modified by subsequent changes of the semifluid mud; and hence the distinctive marks of organization disappear, each toe being simply represented by a linear depression, that has sometimes been mistaken for the impress of a slender toe (Pl. 20). All tridactylous footprints, or those ascribed to birds, that do not exhibit the phalangeal divisions of the toes, have been modified by disturbing causes. It was not until my attention was directed to the localities at Turner's Falls in 1842, that examples of footprints were procured that presented the phalangeal outlines of the toes. The perfection of these footprints proves unequivocally that the feet of the animals making them were distinguished relatively by massive toes, a stout nail, and broad heel. This is the invariable character of those impressions that preserve structural markings; and such as do not are not reliable. I have sometimes seen footprints in which the divisional lines of the middle toe were visible while those of the lateral toes were not, the flowing of the plastic mud not being sufficient to obliterate entirely the forms of the articulations.

In the imperfect impressions represented by Plate 20, the linear impress of the toes is often repeated upon several contiguous strata, — a phenomenon that is explained upon the supposition that the bird, traversing the unconsolidated strata, penetrated with its foot several layers, and upon withdrawing it, left in each, rude representations of the respective toes. These layers, that are always thin, often multiply as many as five or six linear representations of the same foot. Sometimes there is in the upper layers a prolonged depression, running backward from the heel and joined to it, that is due to the *leg* sinking with the foot, its inclined position causing a projected impression from the heel, corresponding in length to the depth to which the foot itself penetrated. The fourth toe sometimes leaves its impress standing out transversely from that produced by the sinking leg. Such multiplied impressions, although they illustrate the history of the footprints, do not convey accurate ideas of structural forms, and their imperfection excludes their specific arrangement.

The perfect footprints, then, were impressed when the consistence of the stratum was such that it prevented the sinking of the foot, yet yielded sufficiently to take its accurate form; and some of the finest examples therefore are those in which the impress is superficial (Pl. 11). Sometimes the resistance opposed to the weight of the animal produced a flattened impress (Pl. 10, fig. 1). When the drying process had not reached that point, the articular impress is concave and tuberous (Pl. 10, fig. 2), and in proportion as the stratum was softer the impress becomes more imperfect, until it can barely be detected, or it results in mere fissured lines that represent the toes.

Another form of imperfect footprints occasionally happens. If a slight earthy deposit falls upon and adheres to a surface having a footprint, the general aspect of this footprint will be preserved by the adventitious layer, but there will be an entire absence of the finer markings of the foot. If such deposits acquire considerable thickness, all signs of the original impress disappear, and the result is a stratum with a relief cast upon its inferior surface. If also the layer upon which the bird trod be very thin, the impress of its foot may be transmitted to the layer beneath; but the impression thus resulting will not be accurate. It is the stratum upon which the bird steps, and that only, that reflects the true features of the foot: the impress presents a smooth, compressed surface, a clear and unequivocal expression, and sharp and well defined ridges, that separate both the toes and articulation. The perfection of the footprint is in proportion to the smoothness, fineness, stability, and plastic property of the stratum, and when these conditions concur the impress is absolutely perfect, whether of the feet of colossal birds or of insects; while, on the contrary, footprints impressed upon soft mud or beneath water will be so modified as simply to be distinguished by linear depressions that represent the several toes.

If the direction in which the animal moved be parallel to the margin of the water line, there is no essential difference in the character and perfection of the consecutive footprints; but if the direction be transversely, or into or from the water, there is a remarkable difference to be observed among the successive impressions, resulting from variable density of the stratum upon which the animals walked. I have observed instances where the first of a line of footprints was flattened. After a step or two the impressions were tuberous; and, pursuing the line farther, they became imperfect, and finally disappeared altogether. When extensive surfaces are exposed *in situ*, these modifications are always apparent; and it frequently happens that the best and worst specimens of footprints occur together upon the same surface.

Probable sources of the Waters of the Sandstone Basin. — It is an interesting subject of inquiry, to determine whether the waters of the ancient sandstone basin were derived from marine, lacustrine, or fluviatile sources. The topography of the valley of the Connecticut River that corresponds to the area of the sandstone deposition suggests the hypothesis of an estuary or inland sea, connecting with a distant ocean; but, however that might be, it is certain that the level of the waters was not stationary, but was subject to considerable fluctuations. It is unsatisfactory to ascribe this disturbance to tidal influence alone; for under ordinary circumstances it is impossible that the stratum should be sufficiently consolidated in the interval between the ebbing and flowing of tides to retain impressions, and, moreover, as has been stated, the stratum often bears unequivocal evidence of being impressed at distinct and distant intervals of time, while it was acquiring an increasing degree of solidity. This fact cannot be disposed of upon the supposition of tidal agency; it is manifest that the periods of fluctuation were much longer than those occurring between the falling and rising of tides. There is abundant evidence that copious rains watered the earth during the sandstone era; and it is not improbable that the atmosphere was more highly charged with moisture than in the present day. The materials of the sandstone rock are composed exclusively of the detritus of antecedent rocks, which were broken, rounded, or pulverized by some mysterious agency, that could only be moved to the place of their receptacle by the force of powerful currents. The coarser parts were lodged in the remote depths, and constituted the unstratified masses of the sandstone rock, while the finer parts were deposited upon this basis in regular stratification. Hence the presumption is at least plausible, that the accumulated waters of the sandstone basin constituted a narrow estuary or inland sea, having its upper or northern limit at the place now known as Turner's

Falls; and, receiving the turbulent floods of the adjacent regions, its ordinary level was thereby disturbed, while at the same time it received fresh depositions of pebbles, sand, and mud, which, upon the subsidence of the agitated waters, were in turn subjected to the action of solar heat, and became the common rendezvous of multitudes of birds and reptiles.

Classification of the Footprints. — There are numerous difficulties in the way of satisfactory classification of the footprints, owing partly to the unstable condition of the rock at the period of the animal's passage over it, and partly to the presumed fact that footprints of dissimilar magnitude were impressed by identical species during the progressive stages of their growth, and also to the circumstance that many of them present no analogies to the feet of living animals.

As regards the first of these conditions, there are numerous modifications of the footprints, as has already been remarked, that result from the unequal resistance offered by the strata when in a wet condition, or when not sufficiently consolidated to retain the form of the foot. A very large proportion of the footprints have, for this reason, undergone changes that have destroyed their symmetry and structural markings. The modified forms under which the footprints occur are innumerable; and the attempt to reduce such of them as do not present the unequivocal impress of the foot to specific arrangement would be impracticable and useless. When, however, it may be briefly repeated, the stratum was sufficiently consolidated to resist the penetration of the foot, yet yielded slightly to its pressure, a well defined impress resulted, that distinctly retained the rounded forms of the digital articulations, and such only can be subjected to systematic analysis. In well defined examples of footprints, their differences and their individuality may, to a certain extent, be recognized; and these distinctions offer the only ground of classification.

A general resemblance of the footprints not differing essentially except in magnitude, offers a perplexing obstacle to classification. This prominent feature of the footprints can only be explained upon the presumption that the various impressions were due to kindred species. This is illustrated by Plate 2, the figures of which are the footprints of the common hen and chickens. If the same disproportion were found to characterize the fossil footprints, it might be extremely difficult to say whether they were specifically distinct, or were due to the same species.

DESCRIPTION OF THE PLATES.

ERRATUM

Page 41, ninth line from bottom, for *Pl.* 40, read *Pl.* 37.

DESCRIPTION OF THE PLATES.

PLATES No's 1 and 2 are footprints of existing living animals, for comparison.

PLATE I.

Fig. 1, Tracks of a Frog.
Figs. 2 and 3, Tracks of recent small Birds.

PLATE II.

Fig. 1, Track of a Hen.
Figs. 2 and 3, Tracks of Chickens.

PLATE III.

Fig. 1 was described by Dr. Deane in a Communication to the American Academy of Arts and Sciences in 1849, as follows:

This is the smallest example of ornithic footprints known. It is a left foot, and displays the marks of peculiar organization tolerably well. The toes are massive; the inner one showing two lobes distinctly, and the middle and outer ones their respective number, but indistinctly; the nails are also slightly impressed. The impression of the tarsus, or distal extremity of the tarso-metatarsal bone, is perceptible, and, altogether, it is the best example of this most diminutive species I have seen. The step is four inches. This species is rare, and has been found only at Turner's Falls.

Fig. 2 was also described by Dr. Deane in his communication above referred to, as follows:

This is a beautiful specimen of the left and right foot of a bird, probably no larger than the preceding; but the impression differs in this respect, that the toes are more numerous and less massive. The phalangeal ranks are not very distinct; but the general features of the track are very striking. Specimens are rare; I have only seen them at a place called the Race, at the uppermost locality of fossil footprints, four miles above Turner's Falls. This example of footprints is strictly analogous to those of several families of the existing order Passerineæ.

Fig. 3. Of this no description has been given by Dr. Deane. The impressions do not, however, differ essentially from that of figure 1, and may have been made by an individual of the same species, though the footprints are slightly larger and the step considerably longer.

PLATE IV.

Dr. Deane has left no account of these impressions. Figs. 2 and 4 represent, undoubtedly, footprints of the species described by Dr. Hitchcock as the Grallator gracillimus, and Fig. 3, that of the Grallator tenuis, in his great work on the Ichnology of Massachusetts. Fine specimens of the tracks of these species may be seen in the magnificent collection at Amherst, from some of which the figures here given were probably drawn.

The G. gracillimus had a step or stride of about eight inches, and the width of trackway made in walking was about three and one half inches. The step of the G. tenuis was about nine inches, and width of trackway made, two and one half inches.

PLATE V.

This plate was not described by Dr. Deane, though he refers to it, in some general remarks upon the footprints, as exhibiting the impress made by the extremity of the nail of the fourth toe of the bird.

The stone from which this plate was drawn is now in the possession of Roswell Field, Esq., a gentleman who is doing much towards the elucidation of the character of the animals that made the footprints, by his indefatigable perseverance in working out specimens from the localities in his neighborhood, near Turner's Falls, and by his accurate observations upon them.

It would be difficult to give upon paper a finer representation of any object in nature than is afforded, by this plate, of the original stone. The drawing could not be better. The impress of the claw of the fourth, or hind toe, is very perceptible in both the right and left footprint, though the toe itself left no mark.

This species, here arranged by Dr. Deane as of Ornithic character, was probably one of a group of animals described by Dr. Hitchcock as Ornithoid Reptiles, the impressions being those only of the hind feet. Some reference to this and other like cases will be found in the remarks that follow the descriptions here given.

PLATE VI.

Fig. 1. This is not described by Dr. Deane. The stone from which the drawing was made is in the collection at Amherst. The impressions are similar in character to those made by the hind feet of the Apatichnus circumagens of Dr. Hitchcock, and may have been made by an animal of that species. See remarks that follow these descriptions.

Fig. 2. These impressions were described by Dr. Deane in his communication on Fossil Footprints, published in the 4th volume of the Memoirs of the American Academy, New Series. He there says:

This is a beautiful example of footprints, perfectly developed in all respects. The order of articulations, and the imprint of the tarsus and nails are clear and unequivocal. This variety is common at Turner's Falls. The distinctive features of this species consist in the strongly marked and tapering toes, long and blunt nails, the divergence of the lateral toes, and the broad space which separates them from the central one of the tarsus, which is separated from the toes, and is placed further back than usual.

PLATE VII.

The impressions of this Plate are nowhere particularly described by Dr. Deane. Under the classification of the Ichnology of Massachusetts, they must be placed among the many varieties of form presented in the footprints of the species called by Dr. Hitchcock the Brontozoum Sillimanium. The step of this species varied in length from twelve to twenty inches. Width of trackway, about four and one half inches.

PLATE VIII.

Figs. 1 and 2. The same as described by Dr. Deane, and published in the Memoirs of the American Academy, vol. 4, New Series. He there says:

These represent a magnificent species of footprint, which I have selected to illustrate the probability that certain analogous impressions, differing in magnitude, are due to an identical species, being impressed by individuals in various stages of development. The distinctive marks of organization are very striking. The foot is long, narrow, and distinctly impressed. A peculiar mark of distinction in this species is the shortness of the outer toe, projecting backward and forward little more than the inner toe. Another remarkable distinction is the circumstance that the lateral toes embrace and modify the form of the first joint of the central toe. The latter toes diverge less than in other species. The tarsus is invariably impressed, and its lobes and those of the respective joints are exceedingly massive. The nails are perfectly defined. Those terminating the lateral toes have an unusual divergence. The step is very long, in the figures being about two feet. The bird was, doubtless, of great relative magnitude, for the impress is always perfect, a circumstance resulting from condensation by great pressure. These large and powerful birds appear to have outlived other species, for, while those became extinct, or at least disappear, these are found under all circumstances. They abound at Turner's Falls where the strata rest upon igneous rock, and also at the Race, several miles distant, and at intermediate places.

This description should be read having in mind the fact that Dr. Deane, when he wrote it, considered many of the forms presented in the next succeeding eleven plates, as well as those of the one under notice, as impressions of one species, in various stages of growth. That his views in this respect were somewhat modified by subsequent observation, may be judged from the descriptions since given by him of Plates 13 and 15, which follow. It is due to Dr. Deane to state that he did not at any time regard the size alone of an imprint as generally of much value in determining specific character.

The variety of form presented by this plate has been described by Dr. Hitchcock as the footprint of the Grallator cuneatus. This species had a step of from twenty-two to twenty-four inches. Width of trackway, three and one half inches.

PLATES IX. AND X.

These, like the forms presented on Plate 7, are all impressions of the Brontozoum Sillimanium of Hitchcock.

Fine specimens of the footprints of this species, and well defined, are to be found in the collections of Amherst College, Boston Society of Natural History, and of Roswell Field, near Turner's Falls.

PLATE XI.

This presents a beautiful example of the form described by Dr. Hitchcock under the name of Grallator formosus.

The step of this species was twenty-seven inches, width of trackway six inches.

PLATE XII.

The original of this is in the possession of the Boston Society of Natural History. This variety is described by Dr. Hitchcock under the name of Brontozoum validum.

Length of step thirty-three inches, width of trackway six inches.

PLATE XIII.

The footprint upon this plate is distinguished for its massive proportions. Its impress is clearly defined, and presents the outlines and structural markings with great distinctness. It is the left foot of a tridactylous bird, each toe exhibiting the characteristic number of articulations. Its prominent feature consists in the contact of the toes and of the heel, these members being compactly united, but without obliterating dividing lines. The first phalanx of the short or inner toe is joined, partly to the impress of the middle tubercle of the tarsal bone, and partly to the first phalanx of the central toe. The second phalanx of this toe is joined to the first and second phalanges of the middle toe. The first phalanx of the outer or long toe is bounded behind by the tubercle of the tarsal bone that supports it; within, by that supporting the central toe; and the second, third, and fourth phalanges embrace the first and second ranks of the central toe. This toe is relatively short, and is much compressed; the two last phalanges usually being so much blended as to appear to be one, and its close contact to the middle toe imparts to it a curvilinear form. The first and second phalanges of the middle toe are much modified in form by the pressure of the lateral toes. It is relatively long, massive, and fleshy. The tubercle supporting the long or outer toe is large and circular, except at its junction with the first phalanx of this toe; that supporting the central toe is so modified in form by the compression of the first phalanges of the respective toes, as to assume a rectangular form, and that supporting the inner toe does not distinctly appear, as is usually the case. The claws are well marked, broad at their origin, and blunt at their termination. They are not, relatively, long, and such is the rule in all the footprints; they are remarkable for breadth and strength.

The compactness of this footprint seems to indicate that the toes and heel were confined in contact by some membranous union. The impress of that portion of the central toe not joined to the lateral toes is free, and preserves the natural outlines of the articulations. It will be observed that the lines of contact between the heel and first phalanx of the long toe are rectilinear, as is also the junction of this toe with the middle one. All the features of this fine footprint bear the marks of great pressure, the oval curves of the articulations being overcome by it and reduced to straight lines. The impress is not deep, yet, by the great weight of the animal, it is sharply set into the stratum when it was evidently in a stiffened condition.

This variety of footprint occurs more abundantly than any other, and the creature by which it was impressed appears to have flourished throughout the entire period of the sandstone deposition, for its footsteps are widely distributed throughout the strata. This rule applies to all the gigantic ornithoid footprints. They are found at Turner's Falls at the bottom of the stratified series, and thence upward through a thickness of many thousand feet. Wherever the stratified rock is extensively quarried they are sure to be seen, sometimes in groups, or associated with the footprints of other species. I have seen examples of the strata that were impressed with as many as seventy-five distinct footprints, showing series from different individuals, some larger and others smaller than in the plate.

The stride of this powerful bird, if such it be, is about three feet, and its body was consequently elevated upon long legs, as in existing struthious birds. The type of the extinct birds may be assumed to be analogous to some existing brevipennate tribes. It is of course impossible to fix the comparison with satisfactory accuracy, and all that can be said is, that in point of magnitude, massive proportions of the feet, and long stride, there are strong analogies to confirm this belief; and it presumptively follows, that in the habits and instincts of the extinct and living races, corresponding affinities should also exist.

To the above description of Dr. Deane it may be added, that he first described and figured the specimen in the Memoirs of the American Academy of Arts and Sciences, Vol. 4, New Series.

Dr. Hitchcock describes this species under the name of Brontozoum exsertum, in the Ichnology of Massachusetts.

Length of step thirty inches, width of trackway six inches.

PLATE XIV.

No description given by Dr. Deane.

This has generally been regarded as of the same species with that figured on Plate 15; but Dr. Hitchcock has described it as different, under the name of Brontozoum minusculum. Specimens are abundant in the cabinet at Amherst.

Length of step twenty to twenty-seven inches, width of trackway twelve and one half inches.

PLATE XV.

The footprint drawn upon this Plate is that of a gigantic tridactylous bird, that lived in great numbers throughout the sandstone period. These impressions are most abundant at Turner's Falls and South Hadley Falls, and they indicate animals that, for magnitude, are without parallel in modern times. With a single exception, perhaps (Pl. 40), they were doubtless the most powerful of all the animals of this formation.

This enormous footprint is formed upon the true ornithic type, having the requisite number and articulations of the toes. The inner toe has two, the middle three, and the outer four phalanges, exclusive of the terminal phalanges that belong to the claws. The first phalanges of the respective toes are in contact, but the extremities of the lateral toes are considerably separated from the central toe, consequently there is not that modification of form of the toes, caused by contact of these members, that occurs in other large examples (Pl. 13). The tubercle of the tarso-metatarsal bone, supporting the outer or long toe, is broad and semicircular; that of the inner toe is well developed, and that of the middle is compressed and modified by the contact of the first phalanges of the respective toes.

This is a fine example of the colossal footprints. Its impress is superficial, but for that reason its features are correctly reflected. These immense footprints are usually so deeply impressed by the enormous weight of the creature, as to appear much larger than they really are. By sinking into the unconsolidated stratum, the mud is so rolled out in all directions as to give an undue proportion to the footprint. In these cases the impress is rarely or never accurate, being more or less modified by changes that subsequently happened. But the impress of the plate is the true image of the foot. Had it been deeply sunk it would be more prominent to the eye, and its apparent magnitude enhanced, yet it would also be less reliable. As it is, the footprint is immense, and its author has no living successor; there is no bird known with this enormous development of the foot. It is not, however, without its analogies. Certain bones of birds occur in the northern island of New Zealand of a race not remotely extinct, having a foot of equal magnitude and identical in its system of articulations. The Dinornis, an apterous bird fifteen feet high, is no discreditable successor to the sandstone monsters.

These stupendous birds were very numerous throughout the entire period of the sandstone deposition, for their impressions occur at intervals in the ascending series of the stratified portion of the rock, from the bottom to the top. Their powerful organization preserved them from annihilation by their enemies; they had no equals, and they were the rulers of their time.

The footprint above described by Dr. Deane is that of the Brontozoum giganteum of Hitchcock. In the cabinet at Amherst is a slab from Northampton having a row of seven consecutive tracks; and there are many others there of this species.

One of the largest footprints of this huge animal may be seen in the collection of the Boston Society of Natural History.

Length of the step three to five feet, and width of trackway eighteen inches.

PLATE XVI.

This is a photograph, taken directly from the original specimen now in possession of Mr. Roswell Field. It is reduced to one half linear size, and is given because it shows so well the impression made by the integument of the skin upon the plastic surface of the soil.

The species, like that of Plate 14, is the Brontozoum minusculum of Hitchcock.

PLATE XVII.

A photograph of the same species as the last, from a specimen at Amherst, but more reduced, taken to show the impression of the integument of the skin.

PLATE XVIII.

A fine footprint, nowhere particularly described by Dr. Deane, and which appears different from any figured by Dr. Hitchcock in his Ichnology, unless, indeed, it be that of a hind foot of his species the Plesiornis quadrupes. A beautiful specimen of this form and character may be seen in the cabinet at Yale College.

PLATE XIX.

No description by Dr. Deane. The original of this is in possession of Mr. Roswell Field, and the drawings are from two of seven consecutive tracks. These impressions are probably those of the animal described by Dr. Hitchcock under the name of Apatichnus circumagens. See Plate 34 and description, for an account of this species.

PLATE XX.

The several impressions upon this plate are doubtless the footprints of birds, yet they do not sustain an intimate comparison with the other ornithoid footprints illustrated in this paper. The resemblance consists merely in the tridactylous character of the feet. They represent a large proportion of the ornithoid footprints that do not present the articular division of the toes. This class of footprints offers a great variety of modified forms. They are sometimes distinguishable by slender lines merely, and sometimes they are much bent or distorted. They are not actual impressions of the foot, but are changed by having been made upon material which was too soft to retain the genuine impress of the plantar surface of the foot. When the foot is withdrawn after penetrating the semifluid mud, there is a settling or flowing back into the track, that obliterates the phalangeal impressions, and converts the impress of the toes into grooved lines merely. If this variety of impress happens when the strata are thin and fissile, the foot often penetrates several laminæ, and leaves a rude outline of the toes in each, often as many as five or six in succession, and by splitting the strata asunder, each shows a linear impression of the toes, and the lowermost of the series sometimes retains a rude form of the articulation of the toes. Sometimes, too, the upper series exhibits the articulations partially, but as a general rule there is a total absence of the true markings of the foot. In some localities the footprints are ruined in this way, the entire surfaces of the fossil strata being cut up into innumerable impressions. These linear footprints often occur upon strata that are remarkable for perfect impressions also, the first being impressed before, and the latter after, consolidation ensued. Strata often occur with footprints of different degrees of perfection, in proportion as the original clay was soft or hard.

It requires some discrimination to distinguish between the genuine impression and that modified by changes occurring subsequently to the planting of the foot; but if it be recollected that the perfect impression should invariably bear the markings of the joints, claws, etc., no error need be committed on this point. It is not difficult to believe that the examples upon the Plate are due to birds, but it is impossible to prove them to be so. They have the trifid character and general symmetry of the feet of birds, and there the comparison ends. Their imperfection does not admit of specific description. It is not possible to distinguish the innumerable forms that these linear imprints assume. Those delineated upon the Plate are fair specimens of this class, but others occur, so defective as to bear little resemblance to footprints.

PLATES XXI. AND XXII.

These were drawn from the footprints of existing reptiles, and are given for purposes of comparison. Those on Plate 21 represent the tracks of a living alligator.

PLATE XXIII.

In the collection of Mr. Roswell Field, at Turner's Falls, is a stone having nine tracks of an animal undoubtedly of the same species as the one of which footprints are here represented. Other specimens may be seen in the collection at Amherst. Dr. Hitchcock has grouped the animal among the lizards, and given the name of Orthodactylus floriferus to the species.

PLATE XXIV.

Fig. 1. An outline drawing of the figures of this plate was given with an article on the Sandstone Fossils of the Connecticut River by Dr. Deane, published by the American Academy in 1856. He there says:

The impressions have remarkable symmetry. Four thick, tapering toes radiate forward, and in the hind foot the impression of the heel is prolonged backward to a considerable distance, and is broad and flattened. The fore foot is planted a little in advance, and a little outward of the hind foot.

These footprints agree nearly with the description given of the Macropterna divaricans in the Ichnology of Massachusetts; an animal classed by Dr. Hitchcock among the Batrachians.

Fig. 2. The impressions here given, like the preceding, may be found, in outline, upon a plate published by Dr. Deane in the Memoirs of the American Academy, in 1856. He at that time remarked, in relation to it, as follows:

The hind foot consists of three thick, pointed toes, widely spread, and a short toe pointing inward. The heel is projected backward, and terminates in a rounded extremity. The anterior foot is not conformable to the posterior. Four toes, two pointing forward and outward, and two outward and backward.

The species is, possibly, the same as described by Dr. Hitchcock under the name of Xyphopeza triplex, and which he includes among lizards. His description of the generic and specific characters would require that the hind toe of the hind foot should extend backward rather than forward, as our plate shows it to do.

As remarked by Dr. Hitchcock, the animal that made these footprints must have had a body of considerable width, its feet having been placed more than two inches apart in walking.

PLATE XXV.

Fig. 1. Not described by Dr. Deane. The imprints are similar to those of Plate 23, and possibly were made by an animal of the same species.

Fig. 2. This figure is given in Dr. Deane's communication upon the Sandstone Fossils of the Connecticut River to the American Academy, in 1856. He there says:

It is an exceedingly perfect impression of the hind foot of probably a Saurian animal, having a Chirotherian aspect, but the relations of which cannot be determined because of the vestiges of the fore feet not being known.

Notwithstanding the above remarks of Dr. Deane, we think he gave, in the same communication, an outline drawing of the track of a hind foot of the same species, accompanied by that of the fore foot. However this may be, tracks showing impressions of the fore feet have since been discovered, and the animal has been described by Dr. Hitchcock under the name of Anisopus Deweyanus. For consecutive tracks of the same species see Plate 26, fig. 2.

Fig. 3. There is no description of this by Dr. Deane, and the impressions differ in some respects from those of any animal described by Dr. Hitchcock in his Ichnology.

PLATE XXVI.

Fig. 1. Outlines of these impressions, or of some quite similar, were given in Dr. Deane's communication to the American Academy in 1856, and he there says:

They were probably made by a Saurian. The toes have a Chirotherian look, thick and massive. Feet divergent. The fore foot planted near the hind, sometimes partially obliterated by it. Stride very great.

Dr. Hitchcock thinks the animal to have been a Marsupial, and has described it in the Ichnology as such, giving it the name of Anisopus gracilis.

Fig. 2. Outline drawings of the impressions of a fore and hind foot of this species are to be found in Dr. Deane's communication to the American Academy in 1856, and are there given as of an animal probably analogous to the one whose footprints are given in Fig. 1.

This species, like the last, Dr. Hitchcock considers to have been a Marsupial, and he has named it Anisopus Deweyanus.

Fine slabs, with the footprints of both the above species, are to be found in the cabinet at Amherst.

PLATE XXVII.

Fig. 1. The original of this is to be found in the collection of Mr. Roswell Field. The stone has six tracks, but two of which are given. They are of an anomalous character.

Fig. 2. The footprints of this plate are of ornithic character, and, excepting in size, do not differ from those of the Grallator tenuis of Hitchcock. See Plate 4, fig. 3.

PLATE XXVIII.

Fig. 1. This, like the last mentioned, is a representation of the track of apparently a small bird. The place of the original is not now known.

Fig. 2. This is an impression made by a tailed quadruped, possibly the Macropterna vulgaris of Hitchcock. The surface upon which these tracks were made was undoubtedly of soft mud, allowing the feet of the animal to sink deep, and consequently causing a distinct imprint of the tail. The specimen from which the drawing was made is in the collection of Amherst.

PLATE XXIX.

Fig. 1. This track was figured, and a brief account of it given, by Dr. Deane, in the Memoirs of the American Academy, in 1856. It differs in some respects from the description by Dr. Hitchcock of that made by the Macropterna gracilipes, but may be the same.

Fig. 2. A representation of these tracks was also given by Dr. Deane in the Memoirs of the American Academy, in 1856, and were then considered by him to have been made by a Batrachian. Subsequently, in a letter, he mentions them as of a character difficult to determine. They may, possibly, have been made by an animal of the same species as the last, though they differ somewhat in the divarication of the toes and in the width of trackway.

PLATE XXX.

Fig. 1. The impressions of this figure, as well as those of the next mentioned, were supposed by Dr. Deane to have been made by biped tailed animals. They represent but a portion of those upon the original stone, beyond which is a succession of twelve or more footprints. Besides these and the impression made by the tail, the surface is covered over with the markings of rain-drops; and, what is peculiar, the tracks of little streamlets made by a combination of drops. No attempt, of course, was made to represent these last on the drawing, but they render the slab exceedingly interesting.

Dr. Hitchcock has described the species as the Selenichnus breviusculus. He judges the animal to have been a Batrachian.

The stone is in possession of Roswell Field.

Fig. 2. This drawing represents the track of an animal of similar character to the last mentioned, and may be found described by Dr. Hitchcock under the name of Selenichnus falcatus. The impressions of both Figures 1 and 2 of this Plate are of tracks made in soft mud, and it is impossible to determine by them, with certainty, whether they were made by a biped or quadruped.

The quadrupedal impressions delineated upon Plates 31, 32, 33, 34, 35, and 36 are distinctly different from those already described, and constitute a peculiar group. The posterior foot is ornithic in type, that is, it is tridactylous, and conforms in all respects to that of birds. The anterior foot is reptilian in its form and character. The impressions all agree in this particular, but they differ in respect to the relative positions in which they occur, and indicate distinct modes of locomotion by the animals making them. In Plates 31 and 32 the movement is by leaping, in Plates 33 and 34 it is by walking, as in the higher grades of reptiles, and in Plates 35 and 36 by the crawling reptiles. These impressions will repay a careful analysis.

PLATES XXXI. AND XXXII.

The subjects illustrated by these Plates are among the most interesting, and at the same time the most perplexing to comprehend, of all the sandstone vestiges. The footprints are those of quadrupedal reptiles, of a distinct order, having no

CONNECTICUT RIVER SANDSTONE. 49

analogies to the other quadrupedal impressions of the sandstone rock. The points of difference consist in the superior magnitude of the animals making them, and in the obvious fact that the impressions indicate animals whose organization was of a superior grade, and whose movements were of a more complicated character. All other quadrupedal impressions of the sandstone rock suggest animals having their analogies in the existing orders of reptiles; but these are of a character so anomalous, as to forbid comparison with the pedal organs of known reptiles, and, consequently, if they be really due to reptilian animals, they indicate a new and separate order of this class. If the distinctive organizations of the animals, as indicated by the footprints, exclude them from membership in the class of reptiles, an unavoidable inference assigns them to the mammalian class, in which certain types exist that offer, in their pedal systems, more plausible grounds of comparison. The obscure origin of these remarkable footprints, and their intricate combinations, have for a long period rendered investigation difficult and inconclusive; and it is but recently that the full complement of impressions due to each individual has been discovered. A brief narrative of the discovery of the separate prints that constitute the completed series of impressions, will enable the reader to comprehend the difficulties that have embarrassed investigation, and also to advance, by progressive steps, to a comprehension of the suite of impressions that indicate one of the most extraordinary of the sandstone creatures.

In the year 1842, while observing the process of rock cuttings at Turner's Falls, my attention was frequently arrested by solitary footprints, having five radiating toes, that were so perfectly defined as to exhibit their phalangeal system. Subsequently, the impressions were discovered in pairs, planted in opposite directions, as represented at the right of Plate 32. As they were not at first seen in consecutive series, or to be associated with other impressions subsequently developed, their character was enigmatical. In the course of my observations, the oblong concave impressions represented in the lower half of Plate 31, *c c*, which were supposed to be made by the metatarsi of the animal, were developed, and, being associated with the footprints referred to at the right of Plate 32, it was presumed that the creature moved by a succession of leaps, and, dropping upon the ground, rested upon its haunches and anterior feet. This inference was partly correct; but it subsequently appeared that the impressions *c c* of Plate 31 were each connected with a footprint, *b b;* and it was then clearly demonstrated that the animal was a quadruped; that the double impressions on the right of Plate 32 were its anterior, and *b b* of Plate 31 its posterior feet, which were connected, without solution of continuity, with the impressions of the metatarsi, *c c*. From a long range of observations, extending over many years, the resulting conclusions were irresistible that the aggregated impressions were due to a leaping animal, and later discoveries corroborate this belief.

The discovery of another singular feature, that completes the entire group of impressions due to the animal, has been but recently made; and that is, the impress of the terminal extremity of the vertebral column, or of the truncated *os coccygis* (Pl. 31, fig. *d*). The impress is so excellent that its character is unmistakable, and it reveals a structural organization of the animal that is, perhaps, without existing analogies. Neither, then, is Plate 31, or 32, separately, complete; but combined, they, include the aggregate of impressions made by the animal when alighting upon the earth, by leaping. Plate 31 lacks the impress of the anterior feet, and Plate 32 that of the coccyx. It will be interesting to examine these several impressions in detail.

The anterior feet, Plate 32, are constituted each of five massive radiating toes. The central one is largest, and is divided into four phalanges, the two contiguous ones into three, and the two lateral ones into two, each. The carpus that supports the toes does not leave its impress, consequently the footprint is digitigrade. These anterior footprints do not differ essentially, in general form and arrangement, from the corresponding feet of certain other sandstone reptiles; the number and form of the toes is identical, the principal point of difference consisting in magnitude.

The posterior feet are upon a dissimilar plan, and agree accurately with the bipedal tridactylous footprints. Like the footprints of birds, the inner toe has two, the middle three, and the outer four, phalanges, and each toe is terminated by a stout nail. So far, then, as the number, form, and arrangement of the toes are concerned, there is no difference, however immaterial, between them and those of birds. But the impress of the tarsus, joined in an unbroken piece with that of the foot, is a feature that never exists in the ornithic footprints. It was remarked elsewhere, that when the foot of the bird penetrated deeply into the unconsolidated stratum, the inclined position of the leg sinking with the foot left a projecting line running backward from the foot, corresponding in length to the depth which the foot itself sunk. But the impress, under these circumstances, is invariably imperfect, whereas the impress of the tarsus, represented by the Plate, is definite and, like that of the foot, unblemished. Its terminal extremity is oval or rounded, and deeper and broader than the proximal, and the tubercles of bones joining the lateral toes (figs. *c c* of Pl. 31) are distinctly impressed. When the creature alighted upon the ground, it impressed the form of the posterior foot and metatarsus, conjoined.

But the most remarkable feature in this assemblage of impressions is the impress of the *os coccygis* (Pl. 31, fig. D). It is heart-shaped, the apex pointing forward, and its position, in regard to the tarso-pedal impressions, is central; and its flattened or slightly concave impress indicates that the tail, or coccyx, of the creature terminated abruptly, near to its junction with the pelvic bones. That the author of these compound impressions moved by leaping, is not only inferred from their formal arrangement, but is proved conclusively by certain phenomena presented by the photograph (Pl. 31). The momentum or propulsion of the animal was such as to drive it far forward after its posterior feet had touched the earth. The right foot grazed the ground to a considerable extent ere it assumed a state of rest. The first contact of the toes with the ground occurs at *b b*, which mark the furrow or trail produced by these slipping members, and *a a* are similar grooves caused by the nails of the lateral toes. The original specimen is fifteen times greater than the photograph, and the extent of earth really brushed by the toes is more than twelve inches. Another fact concurs to prove that the foot B slipped upon the stratum, in consequence of the momentum by which the animal moved, as is seen in the accumulation of mud anteriorly to the foot, which is gathered into massive ridges. These facts are conclusive, that the foot touched the earth while the animal was moving with velocity and force.

It is difficult to find among existing reptiles the prototype of the creature that impressed these singular footprints. Leaping animals are indeed found in this class, but their feet present no analogies to those under consideration. Among the lower grades of the mammalian class of animals, certain types occur, that offer, in their pedal systems and mode of locomotion, a plausible basis of comparison, which is to be found in the marsupial tubes. The feet of the kangaroo are conformable, as to the number and arrangement of the toes, and the metatarsus, like that of the fossil impressions, is formed of a single bone, which rests upon the earth with the foot, and leaves with it its conjoined impress. The posterior foot is tridactylous, the middle toe greatly predominating. There are five radiating toes for the anterior foot. In the usual leaping movements of the kangaroo, the posterior feet only are used; but in a state of rest, the anterior feet touch the earth, as they also do in a kind of hopping movement peculiar to the creature, when undisturbed, or when seeking its food. Its tail also rests upon the earth, giving support when sitting, and assisting the muscular action of the legs when making its powerful leaps.

Under all the relations of the footprints to the pedal systems of the kangaroo, the presumption is very strong that the animals making them rank at the bottom of the mammalian class of animals.

To the above description of Dr. Deane may be added, for readier comparison, that the original specimen of Plate 31, which he states to be fifteen times greater than the photograph, shows the impressions of the footmarks to be about four times longer than the representations of them in the Plate. They are, indeed, of about double the magnitude of those given, of natural size, in Plate 32, and which were lithographed from a fine specimen in the possession of T. Leonard, Esq., of Greenfield. Dr. Hitchcock considers the tracks to be those of two species, and has described them respectively, under the names of Anomœpus major and Anomœpus minor.

PLATES XXXIII. AND XXXIV.

The footprints delineated upon Plates 33 and 34 are quadrupedal, and indicate an unknown organization of the animals by which they were impressed. They are allied to the footprints upon Plates 31 and 32, the posterior feet being ornithic, and the anterior feet reptilian. They are distinguished from them by the absence of the tarsal impress, and by a difference of mode by which the animal moved, being in Plates 31 and 32 by leaping, and in those now under consideration by walking. In the impressions upon Plates 33 and 34, there is no appreciable difference between the posterior feet and the feet of certain living birds, the osseous divisions of the toes being identical. The anterior feet, although upon the same plan as those upon Plates 31 and 32, have four toes only. In walking, the anterior foot is placed upon the inner aspect of the posterior foot, and it stands directly outward. In Plate 34 it is in advance of the foot; in Plate 33 it is within and behind; and, in tracing a continuous line of the footprints, the position of the posterior foot is in uniform correspondence with that of a bird, while at the same time it is attended by its companion, as represented by the Plates.

These singular impressions, like those upon Plates 31 and 32, are entitled to particular consideration, inasmuch as they suggest doubts as to an ornithic origin of the bipedal footprints. The posterior trifid feet are just as susceptible of comparison with the feet of living birds, as those bipedal impressions that have been with great reason supposed to belong to birds; but, being associated with other impressions of reptilian character, it is certain that the animal could have no affinities with birds, but belonged to a distinct race.

It is difficult to avoid the presumption that the footprints upon Plates 31, 32, 33, and 34 were impressed by animals that were identical, except so far as they were distinguished by specific difference, and that they could jump like the kangaroo, or walk as quadrupeds, and, perhaps, as bipeds. If the creatures impressing Plates 33 and 34 could move by leaping, they would produce the same series of impressions as upon Plates 31 and 32; and if, on the other hand, the order be reversed, the latter walking as quadrupeds, the impressions would be identical with those of Plates 33 and 34; and there can be no question that, if any of these creatures walked erect, they would produce impressions in no way different from those supposed to be due to birds. From the intimate relations of the two sets of impressions, it is reasonable to suppose that their authors possessed common attributes, and could either walk or jump at pleasure.

If it could be further proved that these animals possessed the power of walking upon their posterior feet alone, the question of the ornithic origin of the sandstone footprints would be definitely settled. If the posterior feet of Plates 33 and 34 were disconnected from their associated reptilian impressions, they would at once, upon the rules of comparison, be pronounced to be ornithic. Prior to the discovery of the ornithic-reptilian impressions, no fact has ever occurred to disturb the theory of the ornithic character of the tridactylous footprints, for the grounds upon which this belief rested seemed impregnable. But a combination of ornithoid and reptilian footprints proves that a portion of the sandstone animals, having ornithic feet, can have no relation to birds; and this fact raises a strong presumption, that the other portion may yet, by future discovery, be determined to belong to quadrupeds.

There is an incidental circumstance, entitled to some consideration in determining the organization of the sandstone animals. Among the multitude of footprints, not one, in thousands of them, presents the impress of the tegumentary papillæ of the toes. In the great number I have studied, I have seen very few examples, and one of them is the photograph Plate 16. However perfect the impression, the pressure of the toes has produced only a smooth, unbroken surface. The absence of this feature is certainly a negative proof against an ornithic origin. It might be supposed that birds so gigantic as those to which these impressions are ascribed, if they were fitted to traverse the land, would possess the same organizations as occur in existing types. The smoothness of the dermoid coverings of the toes would of itself indicate that the animals were constituted for the water rather than the land. In the fine photograph, Plate 16, the style of the dermoid markings does not correspond to that of existing land birds. The integuments appear to be marked in fine lozenge-shaped checks, and not by those round, prominent points that characterize the feet of terrestrial birds. It might, however, be supposed that the impress of such minute bodies would not be retained by the materials of the stratum; but the most delicate objects, as the feet of insects and the minute rays of fishes' fins, are accurately preserved.

To the above remarks by Dr. Deane we will add, that Plate 33 represents the species described by Dr. Hitchcock as the Plesiornis quadrupes. Plate 34, figs. 1 and 2, represent impressions of the animal described by Dr. Hitchcock as the Apatichnus circumagens. The specimens are in the collection of Roswell Field, Esq.

PLATES XXXV. AND XXXVI.

Like the footprints upon the four preceding Plates, those upon Plate 35 are also of compound character, tridactylous behind and reptilian before. They indicate animals of unknown type. It is evident that they moved by crawling. The posterior foot is analogous to the feet of some living birds, and does not differ from those sandstone footprints that have been described as due to birds. The anterior foot has four toes, and is reptilian in its character. It points outward, and falls, in walking, a little in advance of the posterior foot. The relative positions of the right and left feet vary essentially from those of the preceding Plates. In Plates 33 and 34, the feet fall, in walking, in a direct line nearly, while the right and left feet, in Plate 35, are broadly separated. All the feet drag upon the ground, and the body of the creature was therefore but slightly elevated. The position and trailing movements of the feet suggest some analogy to existing testudinate types, but the character of the feet forbids this inference. Among the sandstone fossils there are really none that indicate tortoises, although a different opinion has formerly been held. The evidence upon which this inference rests, consists in the frequent occurrence of parallel grooves or furrows (Pl. 36), broadly separated, that have been produced by the feet and legs of an animal moving over soft mud, probably beneath water. The discovery of the fine specimen drawn upon Plate 35 explains the cause of these double rows of furrows (Pl. 36), for it may be supposed, that if the animal to which the impressions (Pl. 35) were due, sank deeply into the soft ground, the dragging of each set of feet would produce the impressions of the deep grooves that have been supposed to indicate the existence of tortoises. Although in these cases the impress of the feet are not preserved, their places are accurately marked, occurring in regular alternation from side to side. If a tortoise were thus to sink, its solid body would plow through the mud and leave a distinct trace, a condition I have never seen.

In the present state of information, it is impossible to comprehend the analogies of the animal making the impressions upon Plate 35.

Dr. Hitchcock describes the animal that made the impressions represented on Plate 35 under the name of Tarsodactylus caudatus. The impressions figured on Plate 36, he refers to a species of his genus Helcura.

The original of Plate 35 is in possession of Amherst College.

PLATE XXXVII.

The footprint that forms the subject of this Plate is remarkable for magnitude, being seventeen inches in length and eleven in breadth. It indicates the most colossal of all the sandstone animals. Tridactylous footprints occur of equal length, but for solid, massive proportions, this is unequalled. The impress is very perfect, and shows the osseous divisions of the foot. There are four ponderous toes, joined in contact to the heel. They do not materially differ in length; the two central are longest, and project further forward than the lateral, which, in turn, project furthest behind. They are symmetrically arranged, are in contact, and are nearly parallel. The central toes are divided into five articulations, and the lateral into four each, and neither are surmounted with claws. The heel is distinctly impressed; it is massive, broad, and nearly as long as the toes. The stride is about three and one half feet, and the body of the animal was therefore considerably elevated. These are the prominent features of this remarkable footprint.

There is no fore foot accompanying it, and the inference is, that the creature was a biped. It is difficult to say which the impression resembles most, the footprint of a bird or reptile. The style of locomotion is that of a bird, but the structure of the foot does not conform with that of birds. Neither does it conform with that of reptiles, and has no known analogies to any living animals. The organizations and instincts of the creature cannot, therefore, be comprehended. It was probably some gigantic reptile that very rarely visited the grounds upon which the other sandstone animals congregated. The broad, clawless feet suggest the hypothesis that these organs might be used for propulsion in water, as well as in walking upon land, and that the animal might be some enormous amphibian. Its organization was peculiar to its time. In the transition period of the sandstone deposition, there doubtless existed animals whose organizations have not been transmitted to succeeding ages. All paleozoic eras have been identified by animals peculiar to each; and in this respect the sandstone epoch forms no exception. It was an era replete with wonderful beings, and the combinations and varieties of the living organisms were truly amazing. The character and habits of a portion of the animals may be inferred with confidence; but that of another portion, among which is the monster indicated by the footprint of this Plate, is veiled in profound mystery.

Since the above remarks were written by Dr. Deane, the huge animal that made the tracks has been shown to have been quadrupedal, by the discovery of impressions of the fore feet. Dr. Hitchcock has described it in his great work on the Ichnology of Massachusetts, as the Otozoum Moodii. He arranges it under the order of Batrachians, but thinks it has, "combined in its nature, characteristics now distributed among several different families of animals."

PLATE XXXVIII.

Fig. 1. Drawings in outline of the tracks here represented were given by Dr. Deane in his communication to the American Academy, in 1856. He then supposed them to have been made by a Batrachian, the impressions of the anterior feet not being retained. Subsequently, he judged them to have been made by a biped.

Dr. Hitchcock thinks the impressions to be Chelonian, and has described them in the Ichnology as of an animal which he calls the Exocampe ornata.

Fig. 2. Drawings in outline of these tracks were also given by Dr. Deane, in the communication above referred to. Like the last mentioned, he judged them to have been made by a Batrachian.

Dr. Hitchcock ascribes them to an animal belonging to the group of lizards, which he calls the Orthodactylus linearis.

The specimen here represented is in the cabinet of Amherst.

PLATE XXXIX.

The impressions shown on this Plate, Dr. Deane supposed to have been made by a bipedal reptile. The stone is in the collection of Amherst. In the Ichnology of Massachusetts, the animal is classed among the Chelonians, and has received from Dr. Hitchcock the name of Exocampe arcta.

PLATE XL.

This Plate and the next succeeding present a series of photographic delineations of the tracks of insects, or possibly of small crustaceans. Of all the impressions upon the sandstone of the Connecticut Valley, perhaps none have excited more astonishment upon the minds of beholders, than have those of which figures are here given. The perfect portrayal of the original stones presented by these Plates is remarkable.

Dr. Deane gave some account of these impressions in his communication to the American Academy, in 1856, accompanying it with some Plates; and Dr. Hitchcock has since named, and more fully described them, in his Ichnology of Massachusetts.

The names applied by him, as far as recognized, are here given.

Fig. 1. Acanthichnus saltatorius; specimen in the collection of Roswell Field.

Fig. 2. Conopsoides larvalis (?), from a specimen in the possession of Amherst College.

Fig. 3. Acanthichnus tardigradus, from a specimen in the possession of Amherst College.

Fig. 4. (?). From a specimen in the collection of Amherst College.

Fig. 5. Hexapodichnus magnus; specimen in the collection of Roswell Field.

PLATE XLI.

Fig. 1. Lithographus hieroglyphicus, from a specimen in the collection of Roswell Field.

Fig. 2. Bifurculapes tuberculatus, from specimen in collection of Roswell Field.

Fig. 3. Conopsoides larvalis, from specimen in the collection of Roswell Field.

Fig. 4. Bifurculapes, from specimen in the possession of Wm. Clark, of St. Louis.

Figs. 5 and 6. Bifurculapes laqueatus. Both these figures, taken from different parts of one stone, in the possession of Roswell Field.

PLATE XLII.

Fig. 1. The stone from which this figure was lithographed is in the collection of the late Dr. J. C. Warren, of Boston.

Fig. 2. A part of the stone having this track is in the possession of Roswell Field.

Fig. 3. (?).

PLATE XLIII.

The impression here photographed is one of anomalous character, and Dr. Deane thought it impossible to decipher it.

Dr. Hitchcock thinks it the imprint of some species of fish, to which he gives the name of Ptilichnus anomalus.

The Plate represents the impression the size of the original.

PLATE XLIV.

The four photographs of this Plate represent a continuous impression upon the original stone, now in possession of Roswell Field, and the sections should be joined together in the order they are lettered.

Dr. Deane regarded this as the trail of some animal, made beneath the water. Others have judged it to be of vegetable origin.

PLATE XLV.

Photographs of impressions of recent rain-drops, for comparison.

PLATE XLVI.

Photograph of the impression of rain-drops on the sandstone. Fine specimens are very common in the collections of Amherst, Roswell Field, and the Boston Society of Natural History.

REMARKS.

The student of the preceding pages will have noticed, that among the tracks mentioned in the text of Dr. Deane as made by birds, there are some which have since been classed by Dr. Hitchcock as reptilian. Of these may be instanced Plate 5, and Plate 6, fig. 1, as illustrations. Dr. Deane left no full description of many of the impressions of this volume, but it may be stated, that, at an early period of his observations, he regarded *all* the tracks represented on the Plates, from 3 to 20 inclusive, as unquestionably of ornithic character.

In the progress, however, of his labors, he found reason to distrust his early conclusions, and the writer has a letter written by him to a friend in 1857, in which he says:—

"My investigations, since I commenced, have revealed some remarkable facts, and I am not sure but, in the end, the ornithic doctrine of the footprints must be abandoned. Several facts have recently come to light that have a distinct bearing upon this question. Footprints are found associated with a trail, or a fine grooved line running from one foot to another, that cannot be explained upon any supposition other than that the animal had a tail. If this grooved furrow shall be found to have been produced by a tail, it will settle the fact that the impressions were not due to birds. But a still more decisive circumstance lies in the discovery of ornithic footprints and reptilian footprints combined; that is to say, the posterior feet were ornithic, and the anterior reptilian. This extraordinary combination is the first fact that has ever thrown a doubt upon the ornithic origin of the tracks. It is certain that an animal existed having the feet both of birds and reptiles; a quadruped, with anterior feet of five toes each, and posterior feet of three toes. Separated from the reptilian impressions of the fore feet, no one could hesitate, for a moment, to pronounce the posterior imprints ornithic, upon the strict laws of analogy."

Further than this; in some remarks in his Memoir, Dr. Deane implies, that in the contingency of certain discoveries (just such as have been since made), the whole theory of the ornithic character of any of the footprints would be overthrown. It seems proper to quote here this remarkable prediction, now likely, in the opinion of the writer, to be verified. Already has an article appeared in Silliman's Journal sustaining the view that none of the footprints were made by birds; and this by Roswell Field, whose opportunities for observation and study are not surpassed by those of any other man. Dr. Deane, referring to Plates 31 to 34 inclusive, says: "The singular resemblance of the posterior feet, in these remarkable forms of the quadrupedal footprints, to those of ornithic varieties, is a circumstance of significant meaning, associated, as they are, with footprints of unquestionable reptilian type. If it shall be proved by future discoveries that the animals making these complicated impressions possessed the additional power of walking upon their posterior feet alone, the ornithic theory of the footprints would be settled in a summary manner, impregnable as it now seems to be."

The substance of this he again repeats in his description of Plates 33 and 34.

Now, with the purpose of giving to the reader an important fact in relation to the animals that have left their impressions upon the sandstone, as well as to show how much reason Dr. Deane would have found, had he lived, to assume the ground that *none* of the footprints were made by birds, it may be stated that Mr. Roswell Field has now in his possession a slab with a series of tracks upon it, most of which appear to have been made by a bird, as clearly so as any that are found. The footprints follow each other in the requisite order, the right alternating with the left, and they have all the usual ornithic characters in the number of toes impressed, and in the number of phalanges of the several toes. In the progress of the animal over the surface, he seems at one place to have stopped; for the footprints, instead of following each other in the manner of the first, are brought side by side, or nearly so; and here they exhibit themselves, not, however, as before, but as having a long heel on which they rest, precisely as do the marsupials of our day, and as the animals did that made the impressions on Plates 31 and 32. Yet more. Just in advance of these impressions are two others, smaller, and of different character altogether; in fact, impressions of fore feet, showing the animal to have been a quadruped; most likely a reptilian, but possibly a marsupial. Succeeding these last are other tracks, like the first mentioned, showing the animal had resumed an advancing motion.

Now this case is precisely such as supplies the contingency required to settle, in the estimation of Dr. Deane, the question of the ornithic character of the footmarks.

There is one other point which the writer will refer to. Dr. Deane considered some of the impressions to have been made by bipedal reptiles, and so described them, which have since been shown to have been made by quadrupeds. As in the case of the bird tracks, so called, so of these it may be stated, that discoveries have been made since he wrote, which would have changed his views respecting the animals to whom they owed their origin.

In the case of the Otozoum (see Plate 37 and description), a specimen exhibiting the tracks of the fore feet has recently been presented to the Ichnological Cabinet of Amherst.

One reason why there are not full descriptions of all the footprints by Dr. Deane, rather than of a portion of them, may be found in the fact, that he was constantly making new discoveries in relation to them. This led him to delay writing the text for the plates, as he reasonably judged that every additional day's investigation might enable him to make his descriptions more serviceable.

The writer has felt that justice to Dr. Deane required that these remarks should be appended to the descriptions given; and he hopes they may not be found entirely useless, in the further purpose of imparting a little additional matter of interest concerning the foot-prints.

Pl. 1

On Stone from nature by J Deane M.D.   T Sinclair's lith Phila

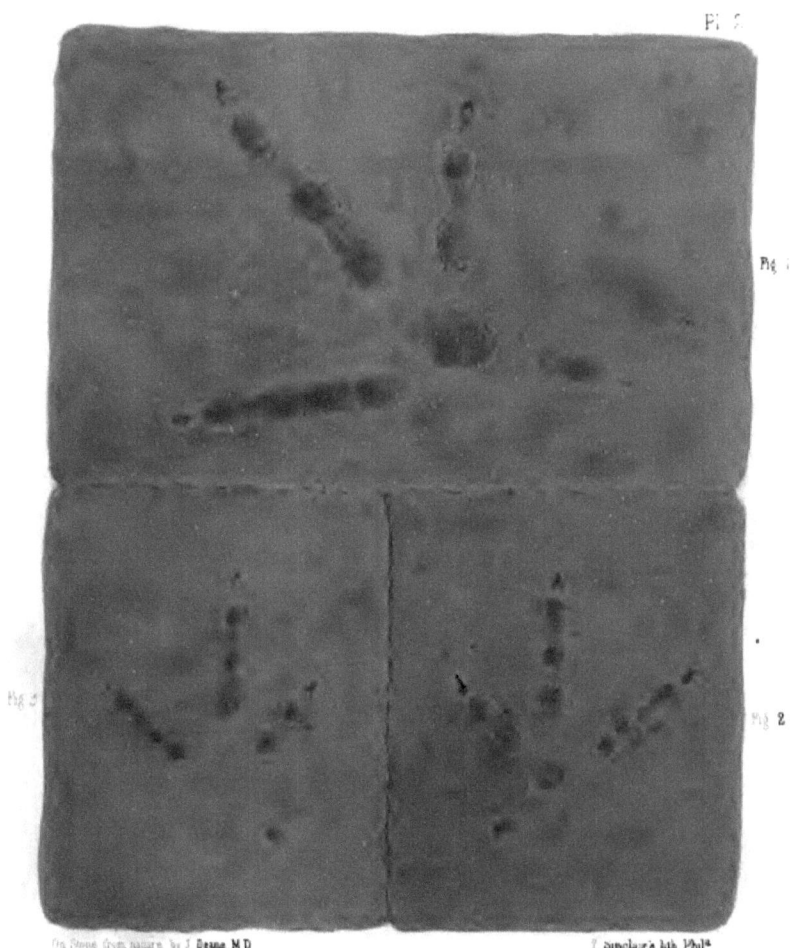

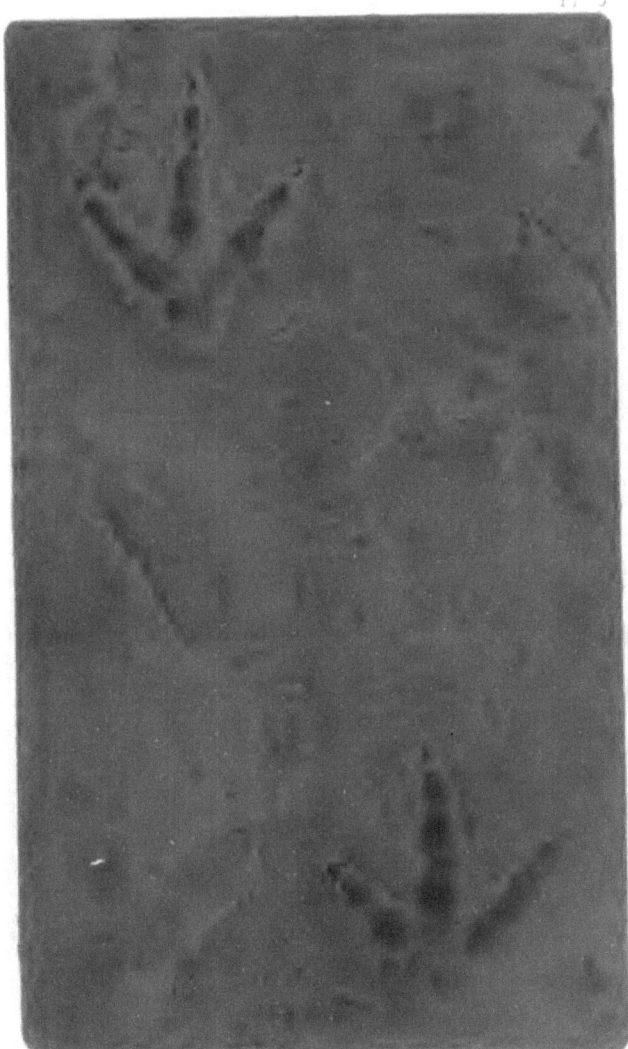

On Stone from nature by J Deane M D   T Sinclair

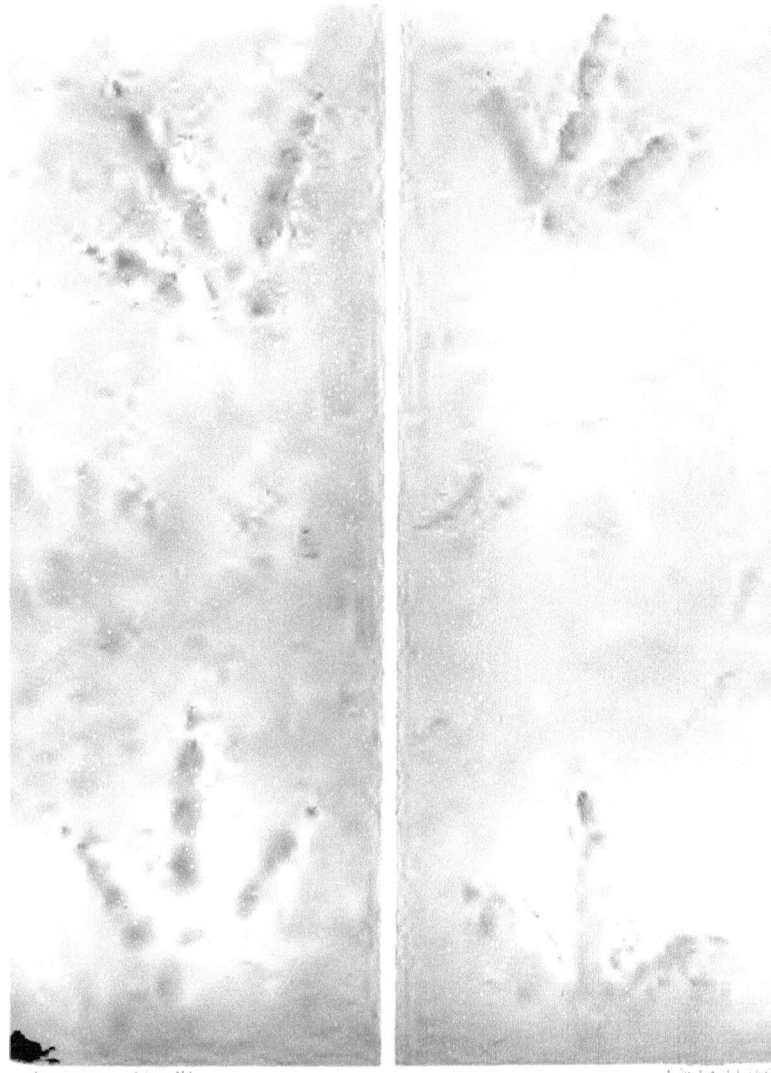

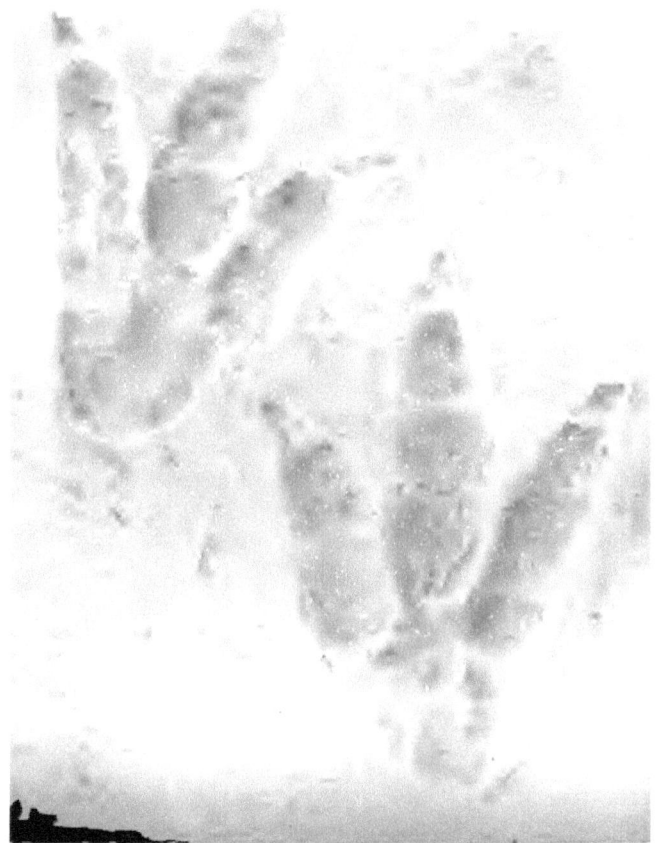

Pl. 8.

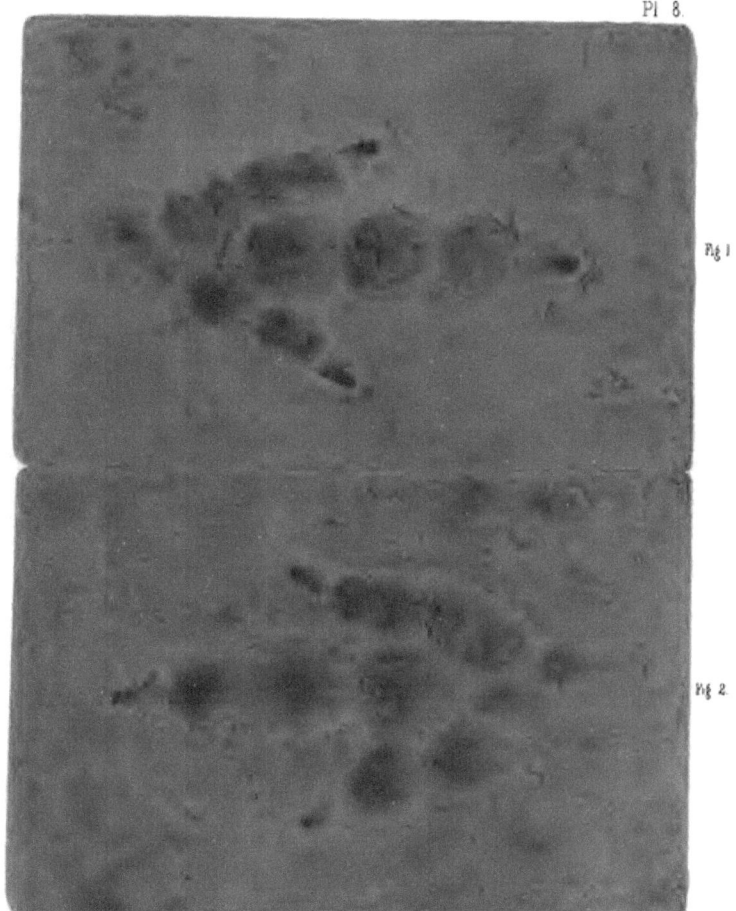

Fig 1.

Fig 2.

On Stone from nature by J. Deane. M.D. 7 Sinclair's lith Phila

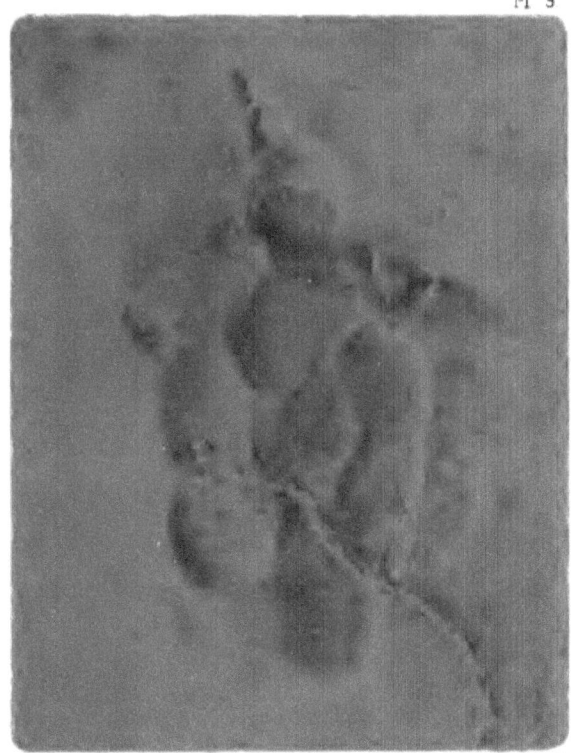

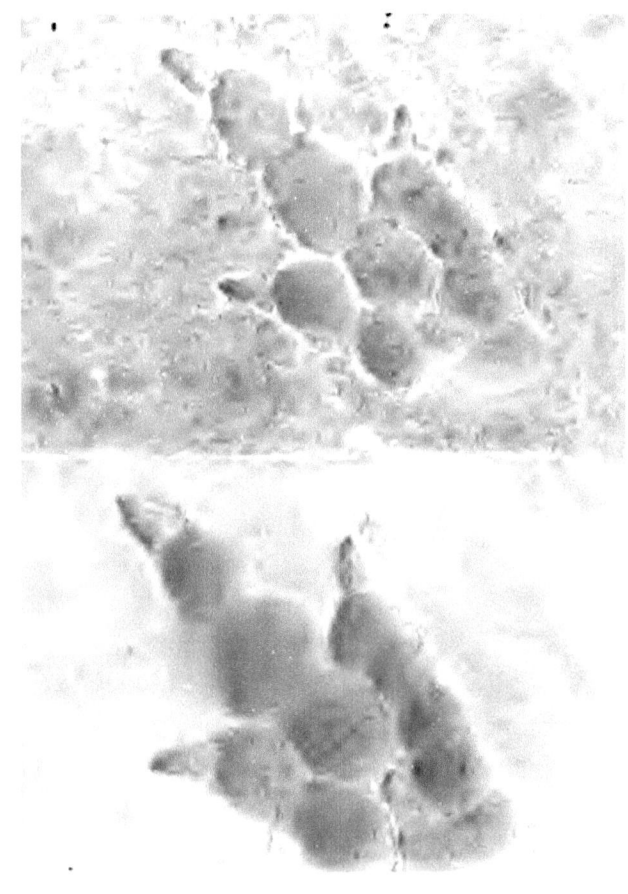

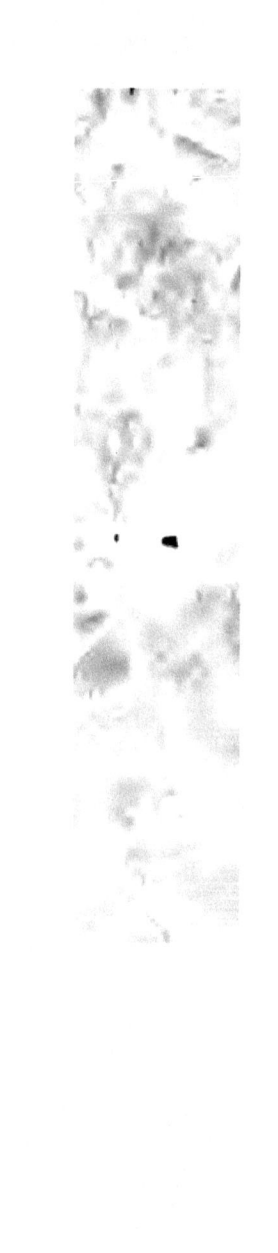

Pl. 16.

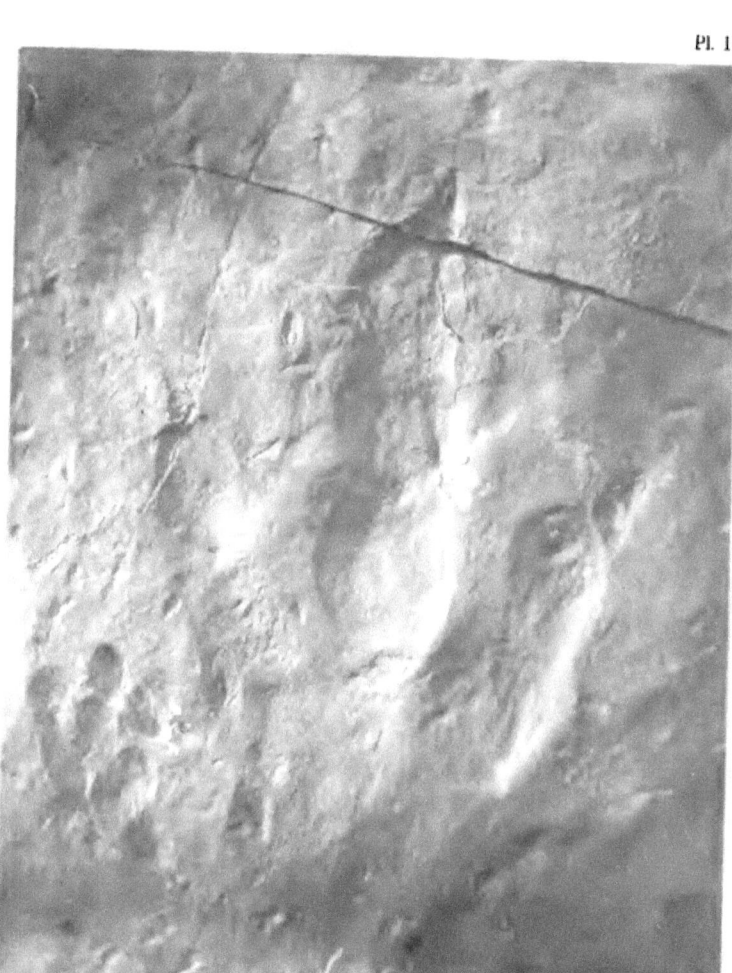

Pl. 17.

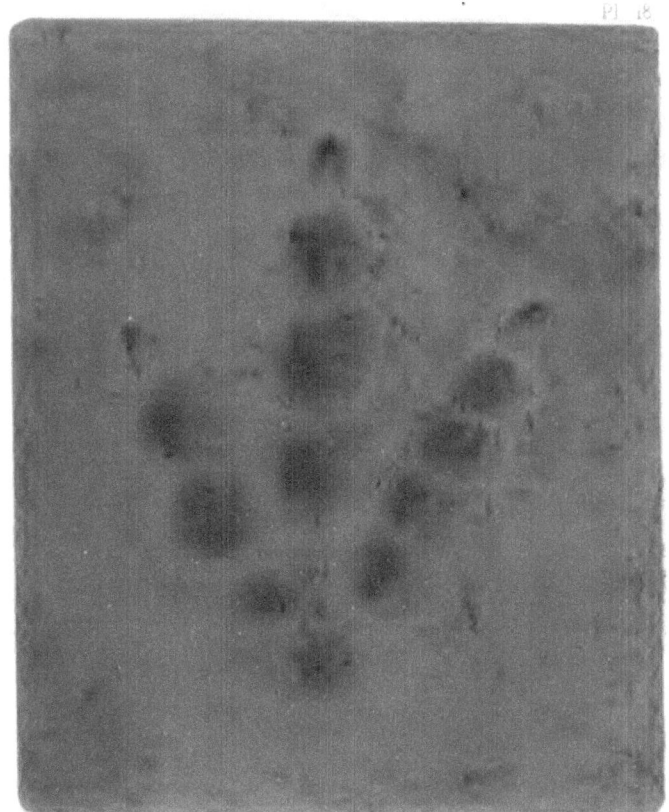

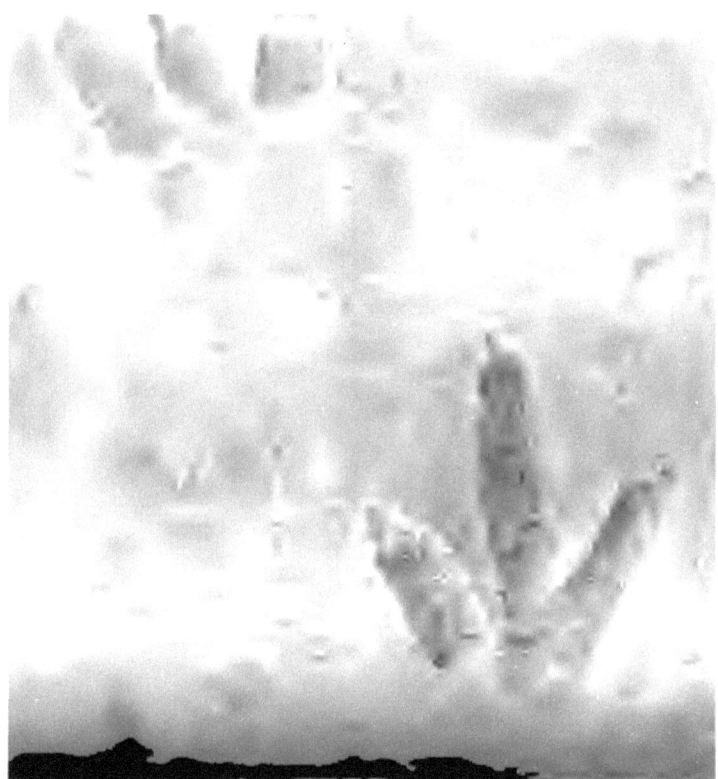

Pl. 20

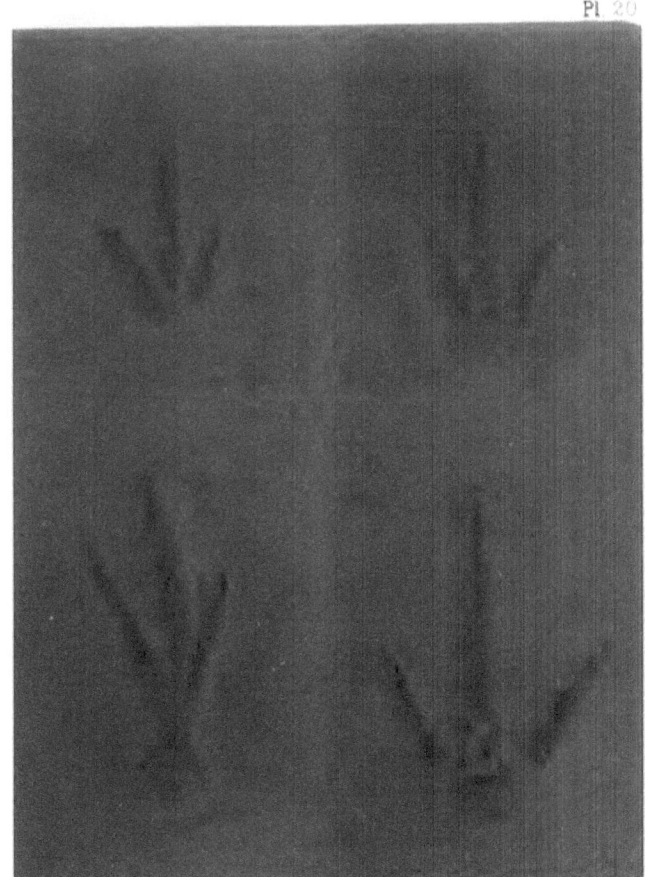

On Stone from nature by J. Deane, M.D.   T. Sinclair's lith. Phila.

Pl. 21

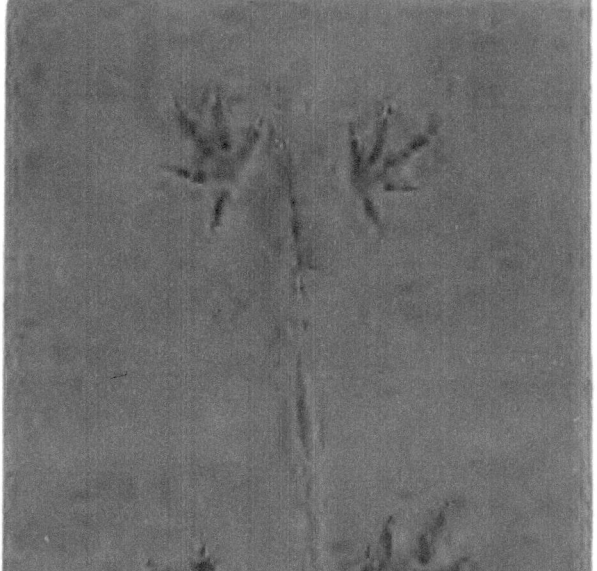

On Stone from nature by J Deane M D.   T Sinclair's lith Phila

Pl 23

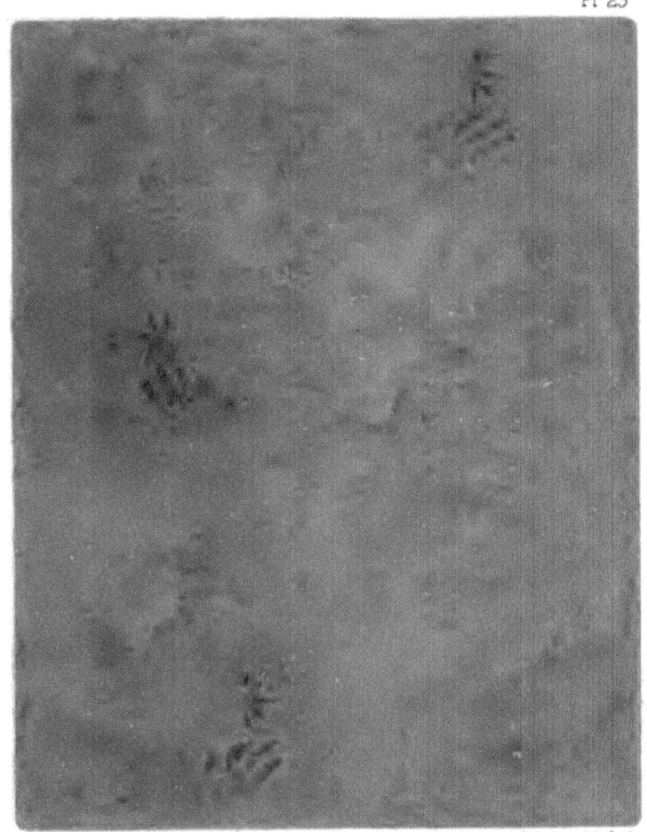

On Stone from nature by J Deane M D. T Sinclairs lith Phila

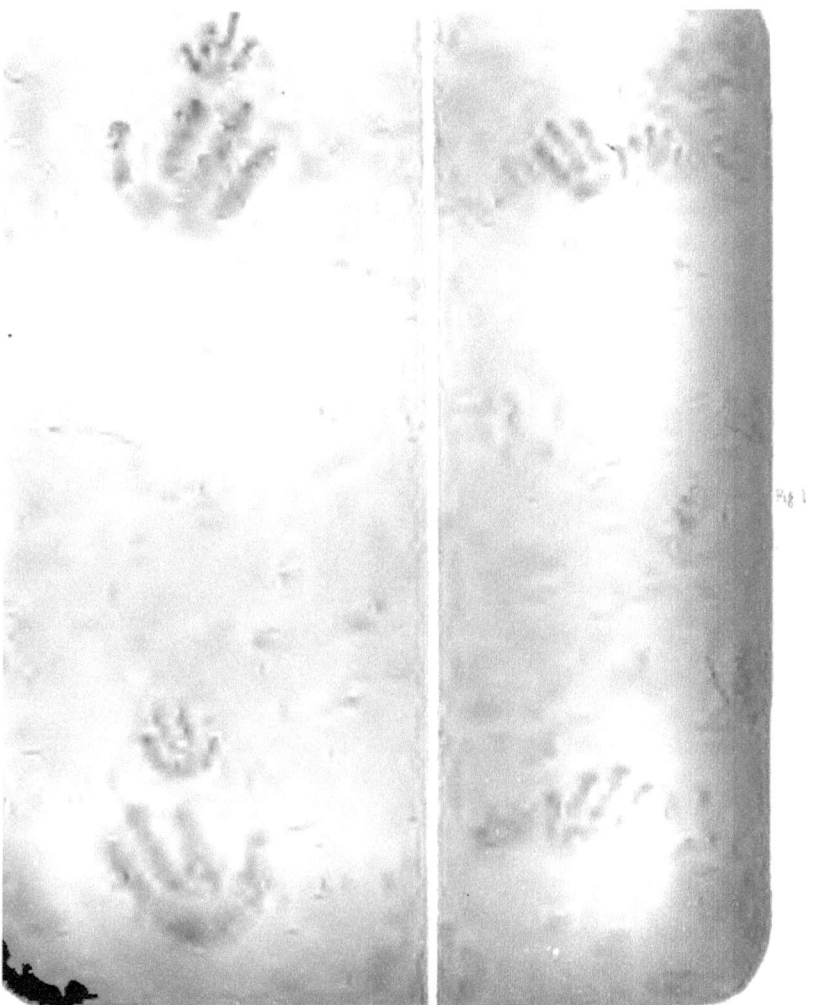

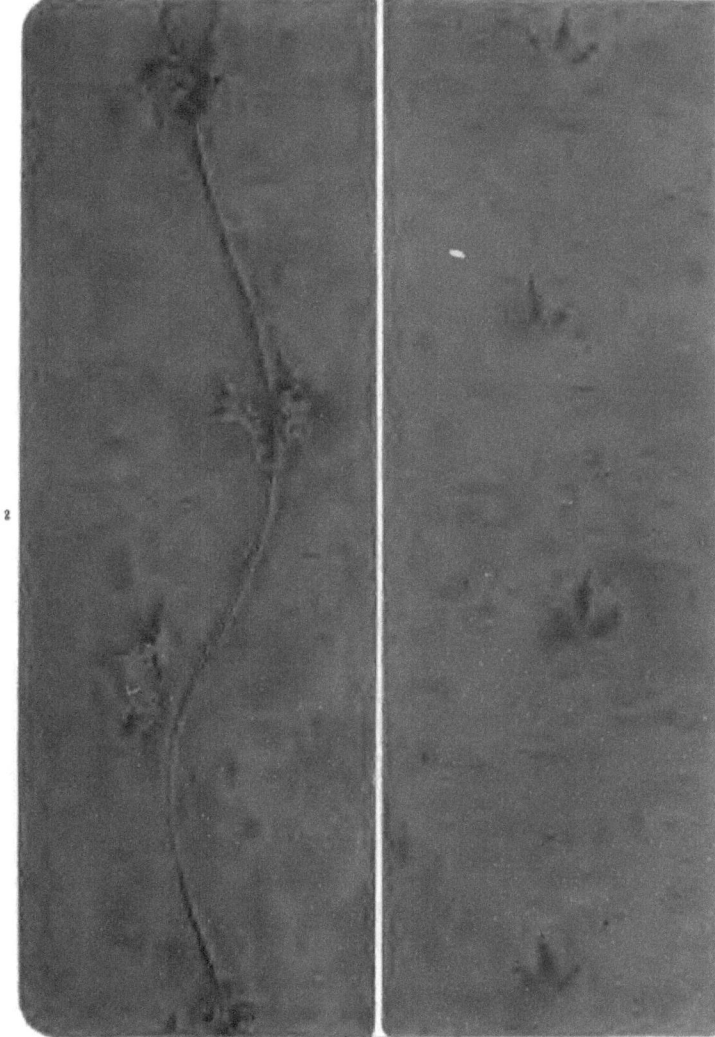

Fig 2

On Stone from nature by J Deane M D F Sinclair's lith. Phila

Fig. 2 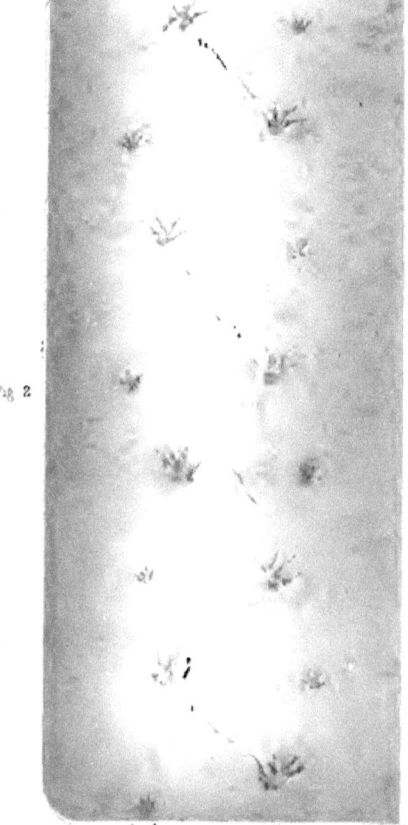 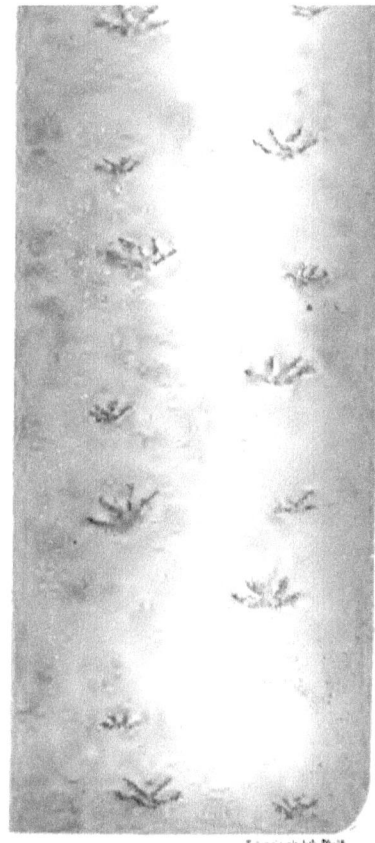 Fig. 1

On Stone from nature by J. Deane, M.D.

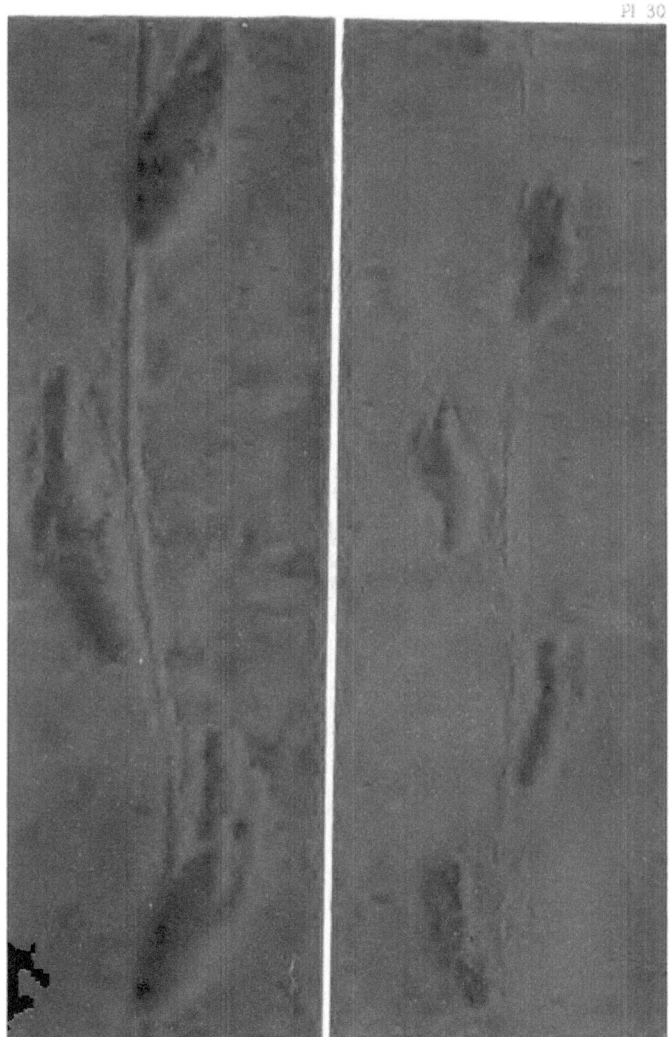

On Stone from nature by J. Deane, M.D. T. Sinclair's lith. Phila.

Pl. 31.

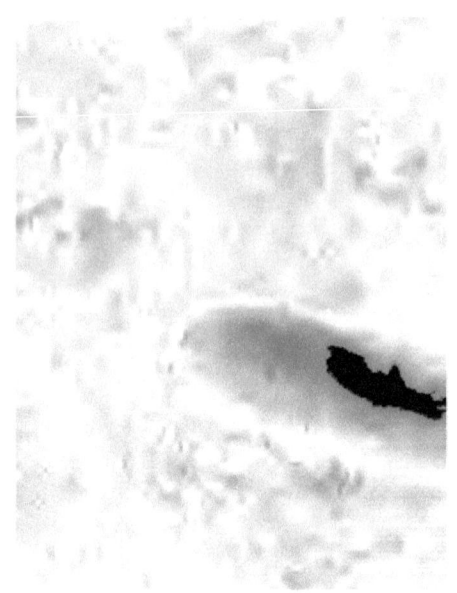

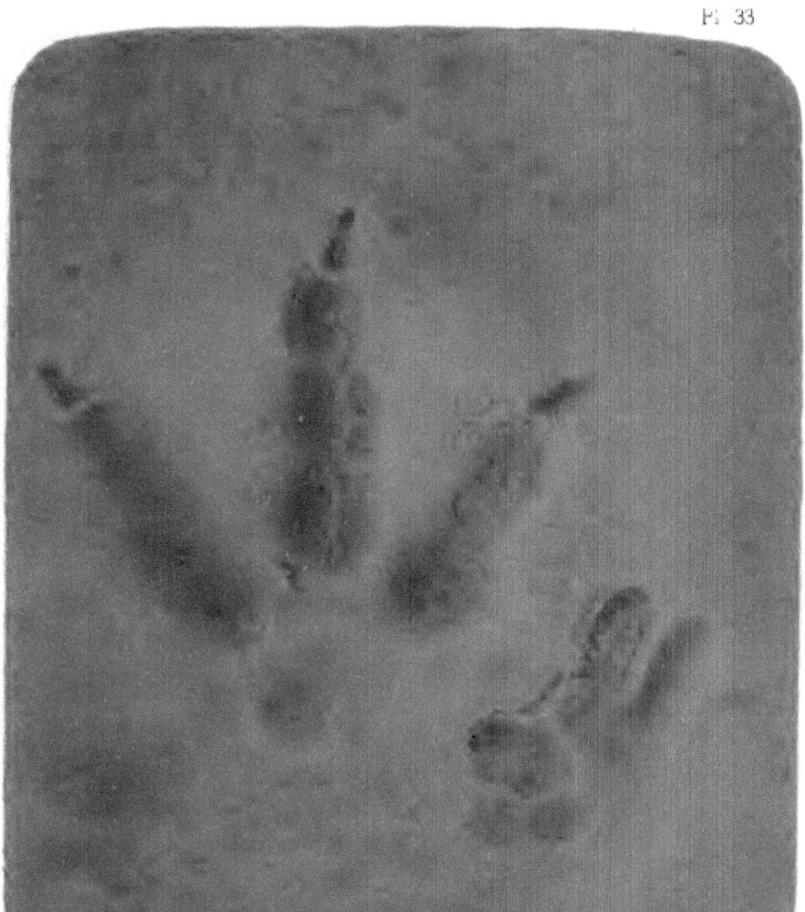

On Stone from nature by J Deane, M.D. T Sinclair's lith Phila

Pl 33

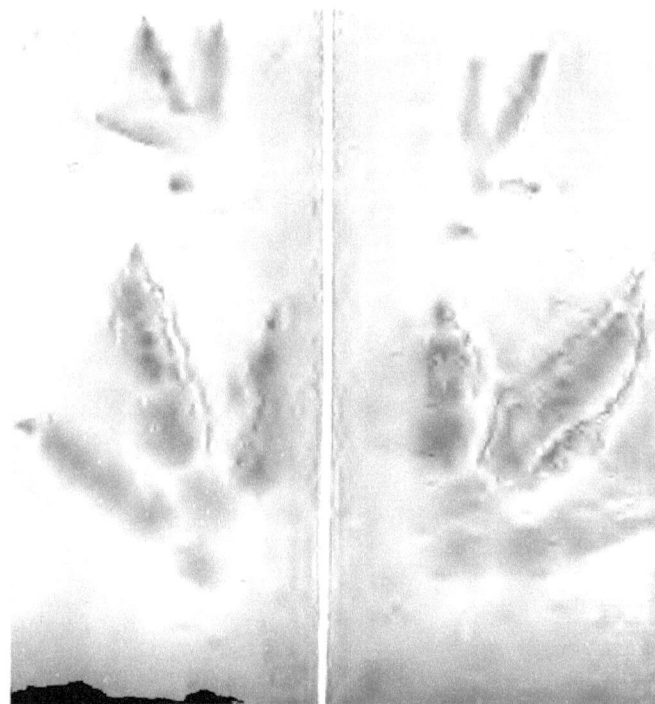

On Stone from nature by J Deane M D

P. S. Duval's lith Phila

Pl 37

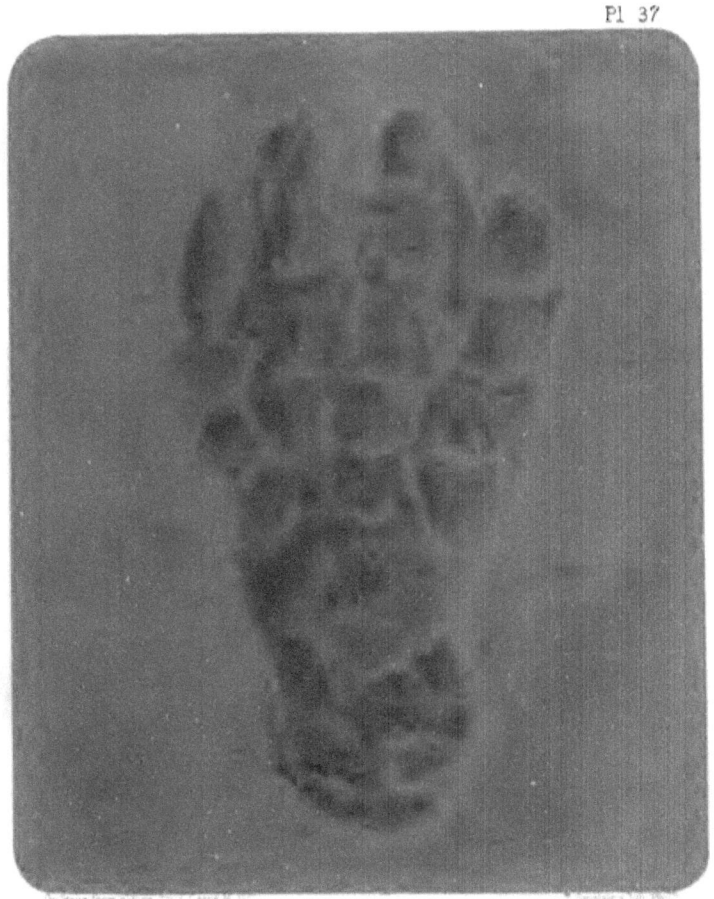

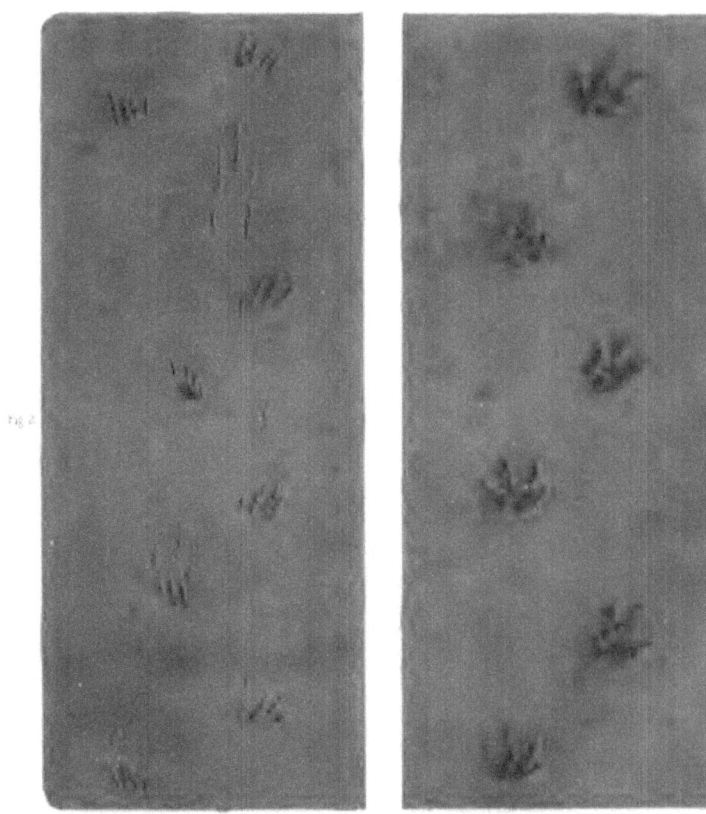

Fig. 2.

On Stone from nature by J. Deane, M.D.

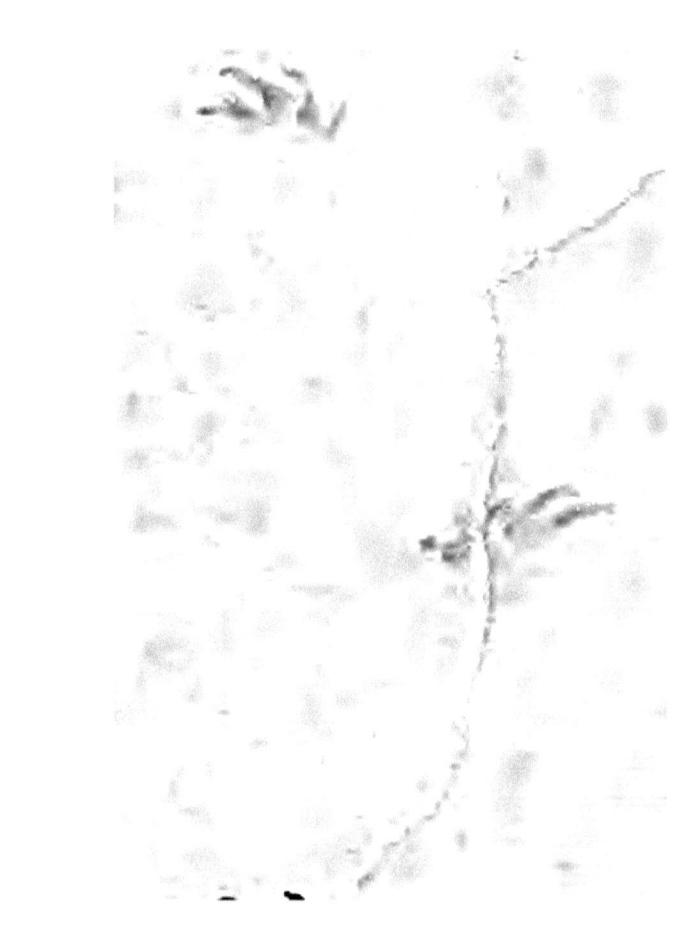

Pl. 40.

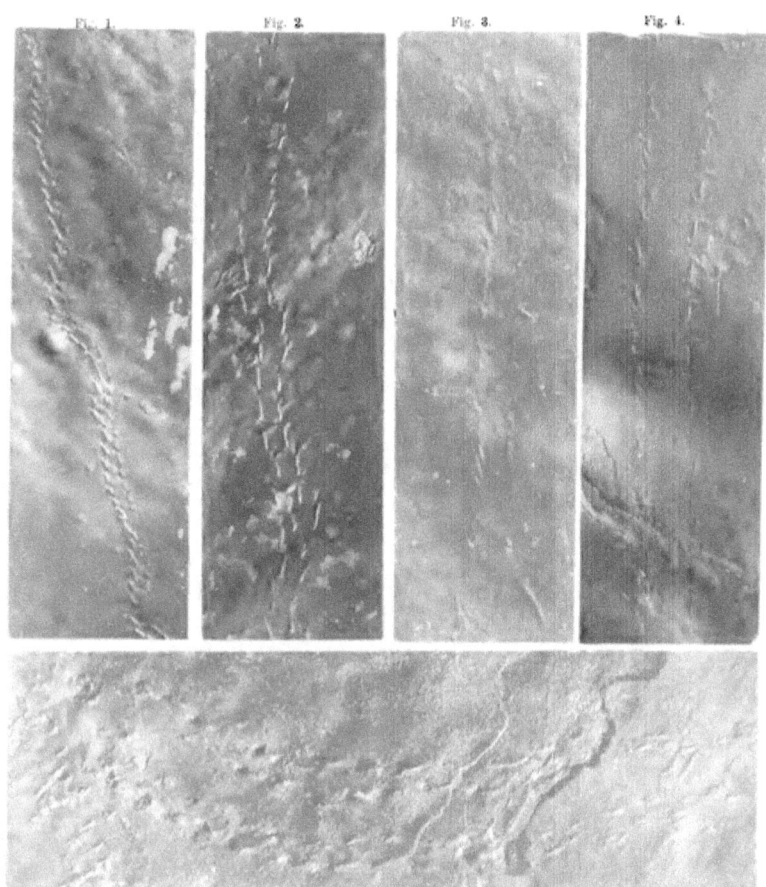

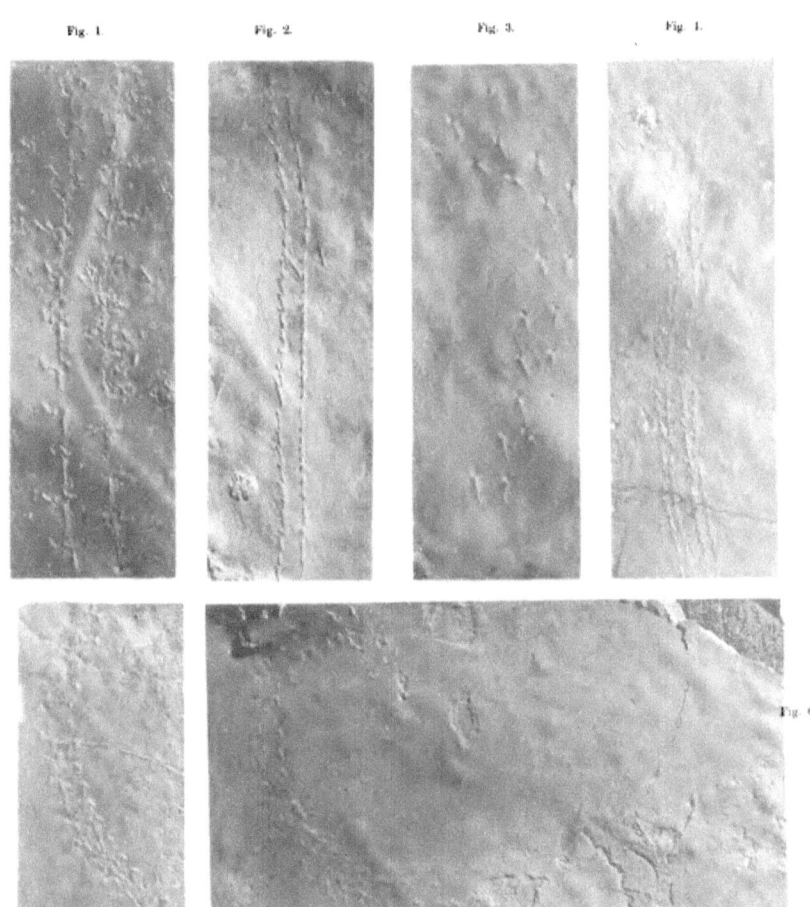

Pl. 41.

Fig. 1. Fig. 2. Fig. 3. Fig. 4.

Fig. 5. Fig. 6.

Stone from nature by J. Deane, M D

Pl. 43.

Pl. 44.

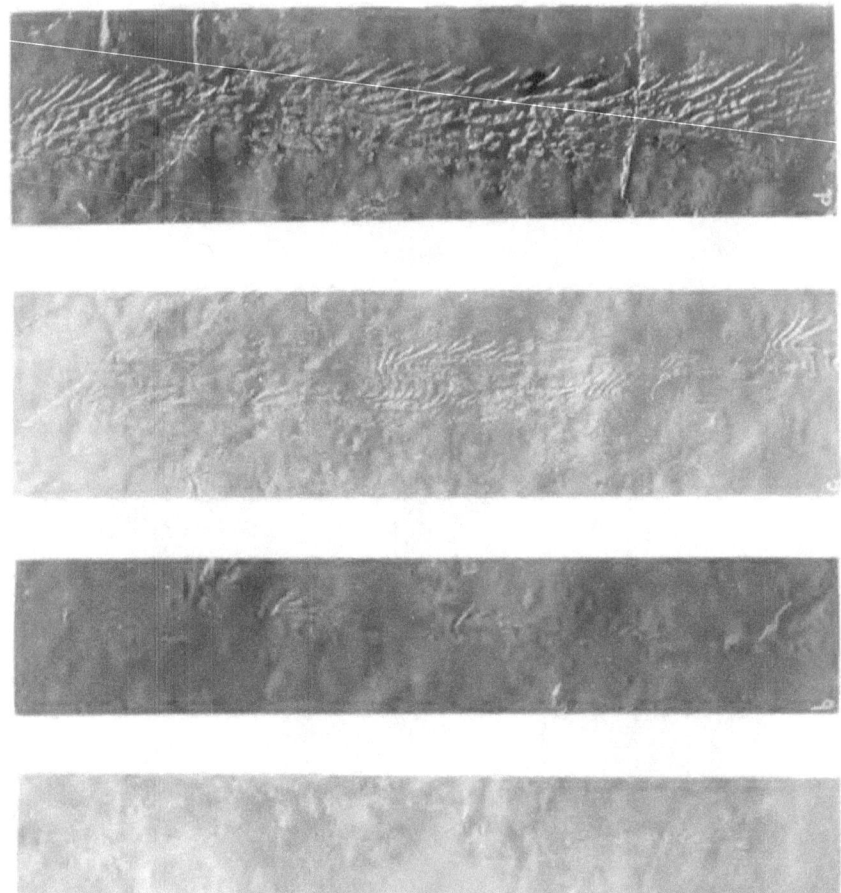

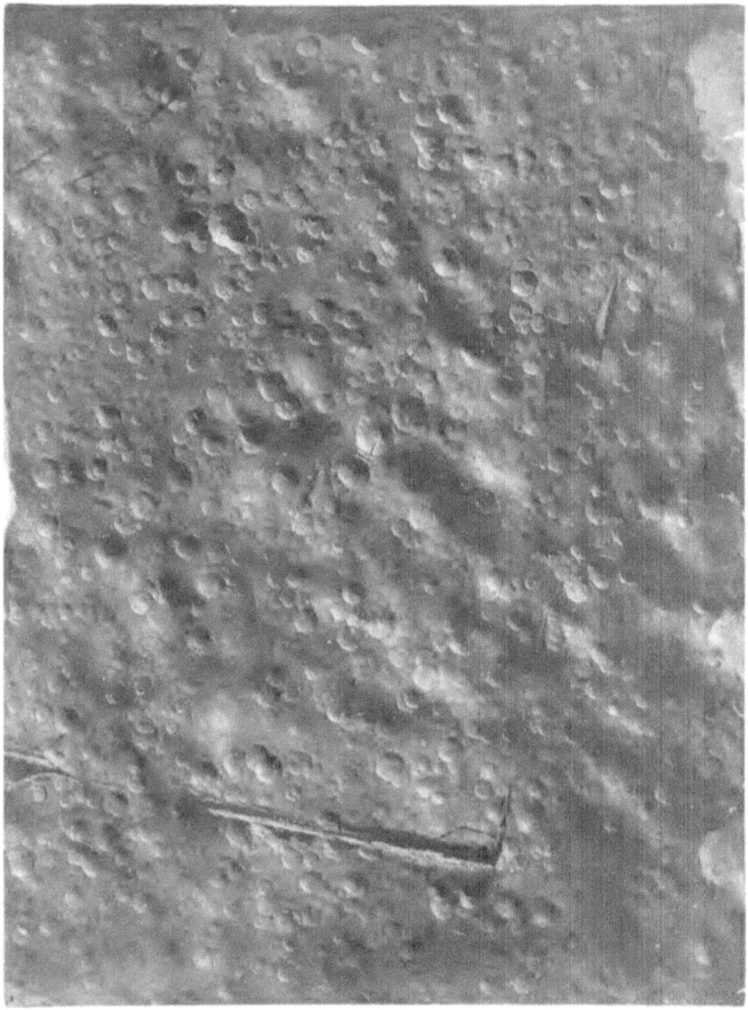

Pl. 46.

www.ingramcontent.com/pod-product-compliance
Lightning Source LLC
Chambersburg PA
CBHW020310170426
43202CB00008B/560